T0181127

Lecture Notes in Computer Science 1347

Edited by G. Goos, J. Hartmanis and J. van Leeuwen

Advisory Board: W. Brauer D. Gries J. Stoer

Springer
Berlin
Heidelberg
New York
Barcelona
Budapest
Hong Kong
London
Milan
Paris
Santa Clara
Singapore
Tokyo

Ehoud Ahronovitz   Christophe Fiorio (Eds.)

Discrete Geometry for Computer Imagery

7th International Workshop, DGCI'97
Montpellier, France, December 3-5, 1997
Proceedings

Springer

Series Editors

Gerhard Goos, Karlsruhe University, Germany

Juris Hartmanis, Cornell University, NY, USA

Jan van Leeuwen, Utrecht University, The Netherlands

Volume Editors

Ehoud Ahronovitz
Christophe Fiorio
LIRMM
161 Rue Ada, F-34392 Montpellier, Cedex 5, France
E-mail: {aro/fiorio}@lirmm.fr

Cataloging-in-Publication data applied for

Die Deutsche Bibliothek - CIP-Einheitsaufnahme

Discrete geometry for computer imagery : 7th international
workshop ; proceedings / DGCI '97, Montpellier, France, December
3 - 5, 1997. Ehoud Ahronovitz ; Christophe Fiorio (ed.). - Berlin ;
Heidelberg ; New York ; Barcelona ; Budapest ; Hong Kong ;
London ; Milan ; Paris ; Santa Clara ; Singapore ; Tokyo : Springer,
1997
   (Lecture notes in computer science ; Vol. 1347)
   ISBN 3-540-63884-9

CR Subject Classification (1991): I.3.5, I.4, G.2, I.6.8

ISSN 0302-9743
ISBN 3-540-63884-9 Springer-Verlag Berlin Heidelberg New York

© Springer-Verlag Berlin Heidelberg 1997
Printed in Germany

Typesetting: Camera-ready by author
SPIN 10652671   06/3142 – 5 4 3 2 1 0   Printed on acid-free paper

Preface

DGCI'97 was the seventh in a series of international conferences on *Discrete Geometry for Computer Imagery*. Their aim is to bring together researchers on discrete geometry, image analysis, image synthesis, and discrete models. These proceedings contain the written contributions to the conference, held in Montpellier (France), December 3–5, 1997.

Discrete geometry plays an important and ever-expanding role in the fields of image modeling and image synthesis, as well as image analysis. It deals with **topological** and **geometric** characterizations of digitized objects or digitized images. Related algorithms that compute and preserve these properties are basic tools in well-known problems such as shape representation, image generation or transforms, image analysis or pattern recognition, to mention but a few.

DGCI'97 covers both theoretical subjects and applications. The main topics, closely related to the above listed problems are : topology, geometry, discrete shapes and planes, nD features, 3D surfaces and volumes, models for discrete spaces, image transformation and generation. They are supported by three outstanding contributions from invited speakers:

T.Yung Kong (Queens College, CUNY, USA),

P. Gritzmann (University of Technology, Munich, GERMANY),

P. Lienhardt (University of Poitiers, FRANCE).

We would like to thank all the authors who responded to the call for papers, as well as the invited speakers, all the referees, and the members of the program committee.

We are grateful to the following institutions for their financial support:

- PRC-GDR ISIS (Information, Signal, et Images) and PRC-GDR AMI (Algorithmique, Modèles, et Infographie) joint research programs of the French CNRS (Centre National de la Recherche Scientifique),

- Université Montpellier II,

- Région Languedoc-Roussillon,

- District de Montpellier,

- and the LIRMM laboratory (Laboratoire d'Informatique, Microélectronique et Robotique de Montpellier) for secretarial support.

We also thank all the participants, for lending their moral support to the speakers.

Montpellier, October 1997

Ehoud AHRONOVITZ and Christophe FIORIO

Préface

DGCI'97 était la septième édition de la conférence internationale sur la *Géométrie Discrète pour l'Imagerie*. Cette conférence a pour objectif de rassembler des chercheurs dans les domaines de la géométrie discrète, de l'analyse d'images, de la synthèse d'images ou des modèles discrets. Ces actes contiennent les contributions écrites de la conférence, qui a eu lieu à Montpellier, du 3 au 5 décembre 1997.

La géométrie discrète joue un rôle important en modélisation, en synthèse et en analyse d'images. Elle traite des caractéristiques **topologiques** et **géométriques** d'objets ou d'images numérisés. Les algorithmes sous-jacents qui calculent et préservent ces propriétés sont des outils de base dans des problèmes aussi variés que la génération et la transformation d'images ou la reconnaissance de formes.

Les articles de DGCI'97 recouvrent des sujets tant théoriques que pratiques. Les thèmes essentiels relèvent de domaines comme la topologie, la géométrie, les formes et surfaces discrètes, les caractéristiques d'objets n−dimensionnels, les surfaces et les volumes 3D, les modèles de l'espace discret, les transformations et générations d'images. Trois conférenciers invités ont apporté leur soutien à cette édition :

T.Yung Kong (Queens College, CUNY, USA),

P. Gritzmann (Université de Technologie, Munich, ALLEMAGNE),

P. Lienhardt (Université de Poitiers, FRANCE)

Nous tenons à remercier tous les auteurs qui ont répondu à l'appel à soumission, les conférenciers invités, les relecteurs et les membres du comité de programme.

Les organisations ci-dessous ont contribué par une aide financière et nous leur en sommes extrêmement reconnaissants :

- le PRC-GDR ISIS (Information, Signal, et Images) et le PRC-GDR AMI (Algorithmique, Modèles, et Infographie) du CNRS,
- l'Université Montpellier II,
- la Région Languedoc-Roussillon,
- le District de Montpellier,
- et le LIRMM (Laboratoire d'Informatique, Microélectronique et Robotique de Montpellier) qui a fourni l'aide logistique.

Merci aussi à tous les participants qui, par leur présence, ont apporté le soutien moral indispensable à tout conférencier.

Montpellier, Octobre 1997

Ehoud AHRONOVITZ et Christophe FIORIO

Conference Co-Chairs

Ehoud Ahronovitz	LIRM Montpellier, France
Christophe Fiorio	LIRM Montpellier, France

Organizing Committee

Ehoud Ahronovitz	LIRM Montpellier, France
Christophe Fiorio	LIRM Montpellier, France
Gilles d'Andréa	LIRM Montpellier, France
Jean-Pierre Aubert	LIRM Montpellier, France
Corinne Zicler	LIRM Montpellier, France

Program Committee

Y. Bertrand	CNRS LSIIT, Strasbourg, France
G. Borgefors	Centre for Image Analysis, SLU, Sweden
J.-M. Chassery	TIMC – IMAG Grenoble, France
U. Eckhardt	Universität Hamburg, Germany
J. Françon	ULP Strasbourg, France
A. Kaufman	State Univ. of New York, USA
V. Kovalevsky	TF Berlin, Germany
W. Kropatsch	TU Vienna, Austria
R. Malgouyres	ISMRA Caen, France
M. Mériaux	IRCOM-SIC Poitiers, France
S. Miguet	ERIC, Université Lyon 2, France
A. Montanvert	TIMC – IMAG Grenoble, France
J.-P. Réveillès	LLAIC1, Clermont-Ferrand, France
D. Richard	LLAIC1, Clermont-Ferrand, France
G. Sanniti di Baja	CNR Napoly, Italy
G. Szekely	ETH-Zurich, Switzerland
S. Ubéda	LIP-ENS Lyon, France
M. Tajine	ULP Strasbourg, France
S. Tanimoto	Univ. Washington, USA
P. Wang	Northeastern Univ. Boston, USA

Referees

Y. Bertrand
G. Borgefors
J.-M. Chassery
U. Eckhardt
J. Françon
A. Kaufman
V. Kovalevsky
W. Kropatsch
A. del Lungo
R. Malgouyres
M. Mériaux
S. Miguet
A. Montanvert
J.-P. Réveillès
D. Richard
G. Sanniti di Baja
G. Szekely
S. Ubéda
M. Tajine
S. Tanimoto
P. Wang

Table of contents

Features

From Principles to Applications

Author Index   255

Invited Speakers

Topology-Preserving Deletion of 1's from 2-, 3- and 4-Dimensional Binary Images

T. Yung Kong

Department of Computer Science
Queens College, CUNY
Flushing, NY 11367, U.S.A.

Abstract. In digital topology, a 1 in a binary image is said to be *simple* if its deletion from the image "preserves topology". Two (closely related) sets of necessary and sufficient conditions for a 1 in a 2-, 3- or 4-dimensional binary image to be simple are established. The 4-dimensional cases of these results may be regarded as the principal contribution of this paper. A different discrete characterization of simple 1's in 2-, 3- and 4-dimensional binary images, discovered by A. W. Roscoe and the author, is also presented (without proof).

1 n-Xels and n-Images. Simple n-Xels.

Let n be any positive integer. An *n-xel* q in Euclidean n-space, $\mathbf{R}^n$, is a closed unit n-dimensional (hyper)cube $q \subset \mathbf{R}^n$ whose 2^n vertices have integer coordinates. (More precisely, an *n-xel in $\mathbf{R}^n$* is a Cartesian product of the form $[i_1, i_1 + 1] \times [i_2, i_2 + 1] \times \ldots \times [i_n, i_n + 1]$, where each of the i's is an integer.) In this paper, a *pixel* is a 2-xel in $\mathbf{R}^2$ and a *voxel* is a 3-xel in $\mathbf{R}^3$. We define an *n-dimensional binary image*, or *n-image*, to be a finite set of n-xels in $\mathbf{R}^n$.

An n-image I can of course be represented by a finite n-dimensional array of 1's and 0's in which each 1 represents an n-xel in I and each 0 represents an n-xel that is not in I. For this reason each n-xel in an n-image I may be called a *1 of I*, and each n-xel that is not in I may be called a *0 of I*.

In this paper, we say that an n-xel in an n-image is *simple* if its removal from that image "preserves topology" in the sense of the following definition:

Definition 1. Let q be an n-xel in an n-image I. q is said to be *simple* in I if the polyhedron $\bigcup I$ can be continuously deformed onto the polyhedron $\bigcup(I - \{q\})$ in such a way that throughout the deformation process:

1. All points that were originally in q stay in q.
2. The points in $\bigcup(I - \{q\})$ do not move.

Equivalently, an element q of an n-image I is simple in I if and only if the polyhedron q can be continuously deformed over itself onto $q \cap \bigcup(I - \{q\})$ in such a way that all points in the latter set remain fixed throughout the deformation process. In the language of topology [1], q is simple in I if and only if there is a deformation retraction of q onto $q \cap \bigcup(I - \{q\})$.

This concept is illustrated in Figure 1. When $n = 2$, it is equivalent to the standard concept of an "8-simple" (or "8-deletable") pixel. When $n = 3$, it is equivalent to the standard concept of a "26-simple" voxel.

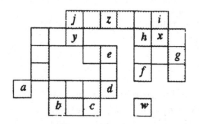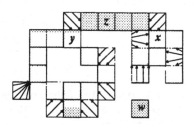

Fig. 1. The left-hand diagram shows ten simple pixels $\{a, b, ..., j\}$ and four nonsimple pixels $\{w, x, y, z\}$ in a 2-image. The right-hand diagram shows a continuous deformation of $\bigcup I$ onto $\bigcup (I - \{q\})$ in accordance with Definition 1 for each $q \in \{a, b, c, ..., j\}$.

Problem 2. How can one determine whether a given n-xel in an n-image I is simple in I?

The problem is trivial in the case $n = 1$. (A 1-xel q in a 1-image I is simple in I if and only if exactly one of the two 1-xels that share an endpoint with q is in I.) There are well known solutions in the cases $n = 2$ and $n = 3$—see section 2. The main goal of this paper is to present a solution in the case $n = 4$.

2 Some Earlier Characterizations of Simple Xels in 2- and 3-Images.

To place our work in context, we first recall solutions to the 2-d and 3-d cases of the above problem.

The following well known characterization of simple pixels solves the 2-d case of the problem:

Theorem 3 (cf. Rosenfeld [7]). *A pixel q in a 2-image I is simple in I if and only if both of the following conditions hold:*

1. *The union of the pixels in $I - \{q\}$ that share at least a vertex with q (i.e., that "are 8-adjacent to q") is nonempty and connected.*
2. *q shares an edge with (i.e., "is 4-adjacent to") a pixel that is not in I.*

It is readily confirmed that the simple pixels $a, b, \ldots, j$ in Figure 1 do indeed satisfy these conditions. The nonsimple pixel x in Figure 1 fails condition 2, while the nonsimple pixels w, y and z fail condition 1.

In the 3-d case, two different solutions to the problem are given by the next two theorems. The first of these theorems uses the concept of the Euler characteristic, which we define in section 3.

Theorem 4 (cf. Tsao and Fu [9]). *A voxel q in a 3-image I is simple in I if and only if both of the following conditions hold:*

1. *The union of the voxels in I − {q} that share at least a vertex with q (i.e., that "are 26-adjacent to q") is connected.*
2. *The union of the voxels in I − {q} that share at least a vertex with q has an Euler characteristic of 1.*

Theorem 5 (cf. Saha et al. [8], Malandain and Bertrand [6]). *A voxel q in a 3-image I is simple in I if and only if the following conditions all hold:*

1. *The union of the voxels in I − {q} that share at least a vertex with q is nonempty and connected.*
2. *q shares a 2-d face with (i.e., "is 6-adjacent to") a voxel that is not in I.*
3. *For every two voxels v and v' that share a 2-d face with q and are not in I, there is a sequence $v = v_0, v_1, \ldots, v_k = v'$ of voxels that are not in I in which each v_i (1 ≤ i ≤ k) shares at least an edge with q (i.e., "is 18-adjacent to q") and shares a 2-d face with v_{i-1}.*

The three theorems above can in fact be deduced from Corollary 8 below without too much work.

Our main theorem (Theorem 7 below) will solve the problem in the case $n = 4$. Another version of the same result (Theorem 9) will give a practical algorithm for determining whether a 4-xel in a 4-image I is simple in I.

3 Xels and Their Faces. Xel Complexes. Boundary(q).

In section 1 we defined the concept of an n-xel in $\mathbf{R}^n$ for $n \geq 1$. More generally, for any integer $k \geq 0$ we define a *k-dimensional xel*, or *k-xel*, to be a closed unit k-cube in a Euclidean space $\mathbf{R}^n$ for some $n \geq \max(k, 1)$. We use the term *xel* to mean a k-xel for some $k \geq 0$. Thus a xel is a Cartesian product $J_1 \times J_2 \times \ldots \times J_n$ where $n \geq 1$ and each J either is a closed unit interval with integer endpoints or is a set consisting of a single integer; the xel is a k-xel if exactly k of the J's are unit intervals. Note that a 0-xel is a singleton set $\{p\}$, where p is a point with integer coordinates.

If X and Y are xels such that $X \subseteq Y$, then we say the xel X is a *face* of the xel Y. A *xel complex* is a finite set K of xels such that every face of each xel in K is itself an element of K. A *subcomplex* of a xel complex K is a set $L \subseteq K$ such that L is itself a xel complex; a *proper* subcomplex of K is a subcomplex of K other than K itself. The *boundary* of a xel q, denoted by Boundary(q), is the xel complex consisting of all the faces of q except q itself. For example, if q is a voxel then Boundary(q) consists of eight 0-xels, twelve 1-xels and six 2-xels.

The *Euler characteristic* of a xel complex K is defined to be the following integer: $\sum_{k=0}^{\infty}(-1)^k$(the number of k-xels in K). There are only finitely many nonzero terms in this sum since xel complexes are finite sets of xels. A well known property of the Euler characteristic is that if K_1 and K_2 are xel complexes such that the polyhedra $\bigcup K_1$ and $\bigcup K_2$ are homeomorphic (i.e., topologically equivalent), then K_1 and K_2 have the same Euler characteristic. If S is any set that is homeomorphic to $\bigcup K$ for some xel complex K, then the *Euler characteristic of S* is defined to be the Euler characteristic of K.

Proposition 6. *Let q be an n-xel. Then the Euler characteristic of Boundary(q) is 2 if n is odd and is 0 if n is even.*

Proof. For all integers $k \geq 0$, let c_k be the number of k-xels in Boundary(q). Now $q = [i_1, i_1 + 1] \times [i_2, i_2 + 1] \times \ldots \times [i_n, i_n + 1]$ for some integers $i_1, i_2, \ldots, i_n$, and each k-xel in Boundary(q) is obtained by replacing just $n - k$ of the intervals $[i_j, i_j + 1]$ with $\{i_j\}$ or $\{i_j + 1\}$. Hence $c_k = 2^{n-k}C(n, n - k) = 2^{n-k}C(n, k)$ for $0 \leq k < n$. Evidently, $c_k = 0$ for $k \geq n$. Thus c_k is the coefficient of t^k in the binomial expansion of $p_n(t) = (2 + t)^n - t^n$. So the Euler characteristic of Boundary(q), which is $\sum_{k=0}^{\infty}(-1)^k c_k$, is equal to $p_n(-1) = 1 - (-1)^n$. $\quad\square$

4 Schlegel Diagrams.

Let n be a positive integer, q an arbitrary n-xel, and K an arbitrary proper subcomplex of Boundary(q).

A Schlegel diagram of K represents K in $\mathbf{R}^{n-1}$. This is particularly useful in the case where q is a 4-xel, because it allows us to think in three dimensions rather than four.

To construct a Schlegel diagram of K, let f be an $(n-1)$-dimensional face of q that is not in K, let a be some point not in q whose orthogonal projection on q is the centroid of the face f, and let P be an $(n-1)$-dimensional hyperplane parallel to f such that q lies between P and the $(n - 1)$-dimensional hyperplane that is the affine hull of f. Consider a projection of $\bigcup(\text{Boundary}(q) - \{f\})$, with a as the "light source", onto P. Let $\pi_{a,P} : \bigcup(\text{Boundary}(q) - \{f\}) \to \mathbf{R}^{n-1}$ be the map induced by this projection. More precisely, for each $x \in \bigcup(\text{Boundary}(q)-\{f\})$ let $g(x)$ be the point at which the line through a and x meets the hyperplane P, and let $\pi_{a,P}(x)$ be the point in $\mathbf{R}^{n-1}$ obtained from the point $g(x)$ by omitting the coordinate whose axis is orthogonal to f and P. Then the set $\{\pi_{a,P}[z] \mid z \in K\}$ will be called a *Schlegel diagram of K*.

The map $\pi_{a,P}$ is a homeomorphism of $\bigcup(\text{Boundary}(q) - \{f\})$ onto its image. Moreover, it is readily confirmed that this map can be extended to a homeomorphism of $\bigcup \text{Boundary}(q)$ onto $\mathbf{R}^{n-1} \cup \{\infty\}$. Here ∞ denotes a single point at infinity. It follows that if S_K is a Schlegel diagram of K then:

(1) $\bigcup S_K$ is homeomorphic to $\bigcup K$.

(2) $(\mathbf{R}^{n-1} \cup \{\infty\}) - \bigcup S_K$ is homeomorphic to $\bigcup \text{Boundary}(q) - \bigcup K$.

(2) implies that $\mathbf{R}^{n-1} - \bigcup S_K$ is connected if and only if $\bigcup \text{Boundary}(q) - \bigcup K$ is connected.

Figure 2 shows a Schlegel diagram of Boundary$(q) - \{f\}$ in the case where q is a voxel and f is one of the six 2-d faces of q. In the case where q is a 4-xel and f is a 3-d face of q, it is not hard to verify that the structure of a Schlegel diagram of Boundary$(q) - \{f\}$ is as shown in Figure 3.

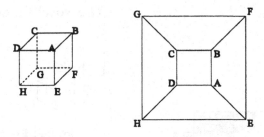

Fig. 2. A voxel q and a Schlegel diagram of Boundary$(q) - \{f\}$, where f is the face of q whose vertices are E, F, G and H.

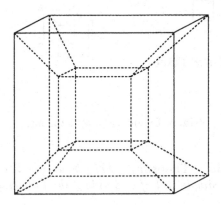

Fig. 3. A Schlegel diagram of Boundary$(q) - \{f\}$ in the case where q is a 4-xel and f is a 3-d face of q.

5   Attach(q, I): The Attachment of an n-Xel q in an n-Image I. The Main Theorem.

Let q be an n-xel in an n-image I. The *attachment* of q in I is the (possibly empty) xel complex Boundary$(q) \cap \bigcup \{$Boundary$(x) \mid x \in I - \{q\}\}$. It is denoted by Attach(q, I). Figure 4 illustrates this concept.

We may think of I as being obtained by "gluing" q onto $\bigcup(I - \{q\})$; in that case $\bigcup$Attach(q, I) is the set of points of $\bigcup$Boundary(q) at which glue may usefully be applied!

Note that $\bigcup$Attach$(q, I) = q \cap \bigcup(I - \{q\})$. Thus q is simple in I if and only if q can be continuously deformed over itself onto $\bigcup$Attach(q, I) in such a way that the points in $\bigcup$Attach(q, I) remain fixed throughout the deformation process.

We are now ready to state the main result of this paper. For $1 \le n \le 4$, the result gives conditions on Attach(q, I) that are necessary and sufficient for an n-xel q to be simple in an n-image I.

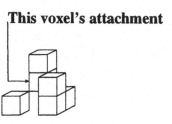

This voxel's attachment

... looks like this:

Here is a Schlegel diagram of the attachment:

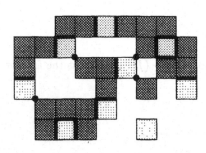

Attachments of Nine Pixels in a 2-Image.

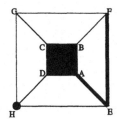

Fig. 4. Examples of attachments.

Theorem 7 (The Main Theorem). *Let q be an n-xel in an n-image I, where $1 \leq n \leq 4$. Then q is simple in I if and only if the following conditions all hold:*

(a) $\bigcup \mathrm{Attach}(q, I)$ *is nonempty and connected.*
(b) $\bigcup \mathrm{Boundary}(q) - \bigcup \mathrm{Attach}(q, I)$ *is nonempty and connected.*
(c) $\bigcup \mathrm{Attach}(q, I)$ *is simply connected.*

Condition (c) of the Main Theorem is needed only when $n = 4$, because (a) and (b) imply (c) when $n \leq 3$. Thus for $n \leq 3$ we have the following simpler characterization of simple n-xels:

Corollary 8. *Let q be an n-xel in an n-image I, where $1 \leq n \leq 3$. Then q is simple in I if and only if both of the following conditions hold:*

(a) $\bigcup \mathrm{Attach}(q, I)$ *is nonempty and connected.*
(b) $\bigcup \mathrm{Boundary}(q) - \bigcup \mathrm{Attach}(q, I)$ *is nonempty and connected.*

This is the same as the characterization of simple 2- and 3-xels given in Theorem 2.10 of [3], though simpleness was defined in a slightly different (but, in fact, equivalent) way in that paper.

Figure 5 shows a violation of condition (b) of Corollary 8. Figure 6 shows a violation of condition (c) of Theorem 7. In both cases q is nonsimple in I. The voxel whose attachment is shown in the right-hand diagram of Figure 4 violates condition (a) of Corollary 8 and is therefore nonsimple.

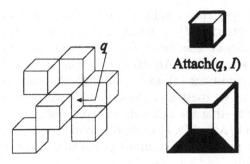

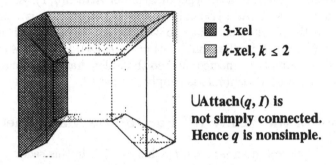

**UBoundary(q) − UAttach(q, I) is disconnected.
Hence q is nonsimple.**

Fig. 5. An example of how Attach(q, I) can violate condition (b) of Corollary 8 in the case $n = 3$.

▓ **3-xel**
░ **k-xel, k ≤ 2**

**UAttach(q, I) is
not simply connected.
Hence q is nonsimple.**

Fig. 6. A Schlegel diagram of an attachment that violates condition (c) of the Main Theorem in the case $n = 4$.

In the following version of the Main Theorem, condition (c) has been changed to a condition on the Euler characteristic of Attach(q, I). As the Euler characteristic is easily computed, this version of the Main Theorem gives a very practical algorithm for determining whether a 1 in a 4-dimensional binary image is simple.

Theorem 9 (The Main Theorem, version 2). *Let q be an n-xel in an n-image I, where $1 \leq n \leq 4$. Then q is simple in I if and only if the following conditions all hold:*

(a) $\bigcup \text{Attach}(q, I)$ *is connected.*
(b) $\bigcup \text{Boundary}(q) - \bigcup \text{Attach}(q, I)$ *is connected.*
(c) The Euler characteristic of $\text{Attach}(q, I)$ *is 1.*

We claim that Theorems 7 and 9 are equivalent. In verifying this we may assume $\bigcup \text{Attach}(q, I)$ and $\bigcup \text{Boundary}(q) - \bigcup \text{Attach}(q, I)$ are both connected; otherwise (a) or (b) fails in both theorems.

If Attach(q, I) is empty then condition (a) of Theorem 7 and condition (c) of Theorem 9 fail. If Attach$(q, I) =$ Boundary(q) then condition (b) of Theorem 7 and condition (c) of Theorem 9 fail—the latter fails by Proposition 6. Thus we may further assume that Attach(q, I) is nonempty and Attach$(q, I) \neq$ Boundary(q), so that conditions (a) and (b) of both theorems hold.

To justify the claim, we must now show that the conditions (c) of the two theorems are equivalent—that the Euler characteristic of Attach(q, I) is 1 if and only if the polyhedron $\bigcup$ Attach(q, I) is simply connected.

It is a well known property of bounded polyhedra in 3-space that for any such polyhedron A

$$\chi(A) = h_0(A) - h_1(A) + h_2(A) \tag{1}$$

where $\chi(A)$ denotes the Euler characteristic of A, $h_0(A)$ denotes the number of connected components of A, $h_2(A)$ denotes the number of 3-d cavities in A, and where $h_1(A)$ (which essentially denotes the number of annulus/doughnut-type holes in A) is 0 if and only if A is simply connected [4].

Let A be the union of a Schlegel diagram of Attach(q, I). As $\bigcup$ Attach(q, I) is connected, so is A. As $\bigcup$ Boundary$(q) - \bigcup$ Attach(q, I) is connected, so is $\mathbf{R}^3 - A$; thus the polyhedron A has no 3-d cavities. So $h_0(A) = 1$ and $h_2(A) = 0$ in equation (1), which implies that $\chi(A) = 1 - h_1(A)$. Hence $\chi(A)$ has value 1 if and only if A is simply connected, and so the Euler characteristic of Attach(q, I) is 1 if and only if $\bigcup$ Attach(q, I) is simply connected.

6   Proof of the "Only If" Part of the Main Theorem.

Let q be a simple xel in an n-image I, where $n \leq 4$. In this section we show that Attach(q, I) must satisfy conditions (a – c) of Theorem 7.

Continuous deformation preserves many topological properties, including connectedness, simple connectedness, and the Euler characteristic. As the n-xel q is an n-cube, it is convex and can therefore be continuously deformed to a point. So, just like a point, q has the following properties:

(q1) q is nonempty and connected.
(q2) q is simply connected.
(q3) The Euler characteristic of q is 1.

As the n-xel q is simple in I, q can be continuously deformed over itself onto $\bigcup$ Attach(q, I). Hence (q1 – q3) imply

(A1) $\bigcup$ Attach(q, I) is nonempty and connected.
(A2) $\bigcup$ Attach(q, I) is simply connected.
(A3) The Euler characteristic of $\bigcup$ Attach(q, I) is 1.

Note that A3 implies $\bigcup$ Boundary$(q) - \bigcup$ Attach(q, I) is nonempty, by Proposition 6.

To prove the "only if" part of the Main Theorem, it remains only to show that $\bigcup$ Boundary$(q) - \bigcup$ Attach(q, I) is connected. We will give an elementary proof of this, based on equation (1) of the previous section.

The result is evidently true if $n = 1$. If $n = 2$, then it follows from A1. If $n = 3$ then it follows from A2. In the case $n = 4$, let A be the union of a Schlegel diagram of Attach(q, I). Then A is connected (by A1), simply connected (by A2), and has an Euler characteristic of 1 (by A3). Thus in equation (1) above we have $h_0(A) = 1$, $h_1(A) = 0$, and $\chi(A) = 1$, so that $1 = 1 - 0 + h_2(A)$, which implies $h_2(A) = 0$. Hence $\mathbf{R}^3 - A$ is connected, and so $\bigcup \text{Boundary}(q) - \bigcup \text{Attach}(q, I)$ is connected, as required.

This argument establishes the "only if" part of the Main Theorem.

7 Proof of the "If" Part of the Main Theorem.

We now prove that if the three conditions of Theorem 7 hold, then q can be continuously deformed onto $\bigcup \text{Attach}(q, I)$. We prove this using a classical discrete deformation process for xel complexes called *collapsing*. As we will see in section 7.2, this process has the property that if a xel complex $K \subseteq \text{Boundary}(q)$ can be collapsed to a "trivial" complex consisting of a single 0-xel, then the xel q can be continuously deformed to $\bigcup K$. We prove that if the three conditions of Theorem 7 all hold then Attach(q, I) can indeed be collapsed to a trivial complex, so that q can be continuously deformed onto $\bigcup \text{Attach}(q, I)$, as required.

7.1 The * Operation. Free Faces of a Complex. Collapsing.

Let q be any xel and let c be the center of q. For every xel x in Boundary(q), we define $c*x$ to be the convex hull of $\{c\} \cup x$; we say that $c*x$ is $(k+1)$-dimensional if x is a k-dimensional xel. For every subcomplex K of Boundary(q), we define $c * K = \{c * x \mid x \in K\} \cup \{c\} \cup K$. For every xel y in Boundary(q), the *faces* of the polyhedron $c * y$ are:

1. The xel y.
2. The elements of $c * \text{Boundary}(y)$.
3. $c * y$ itself.

If $w = c * y$ then the *boundary* of w, denoted by Boundary(w), is the set of all faces of w other than w itself. A subset Z of $c * \text{Boundary}(q)$ is called a *complex* if it satisfies the condition that, for each $z \in Z$, every face of z is also an element of Z.

In the rest of this paper we use the term *complex* to mean either a complex of the kind we have just defined, or a xel complex. (Both of these are special cases of the standard concept of a polyhedral complex, but they are the only cases we will use.)

We now define *collapsing*. This is a standard tool of geometric combinatorial topology.

Let x be a maximal element of a complex K (i.e., an element of K that is not included in any higher-dimensional element of K) and let d be the dimension of x. If $d \geq 1$ then a *free face of x* (in K) is a $(d - 1)$-dimensional face f of x that is not included in any element of $K - \{x, f\}$. A *free face of K* is an element of K that is a free face of z in K for some maximal element z of K.

If f is a free face of an element x of a complex K, then the process of removing both x and f from K is called an *elementary collapse of K, at f, to the complex* $K - \{x, f\}$. A complex K is said to *collapse* to a complex $L \subseteq K$ if there is a (possibly empty) sequence of elementary collapses that transforms K to L. A complex K is said to be *collapsible* if K collapses to a complex that consists only of a single 0-xel. These concepts are illustrated in Figure 7.

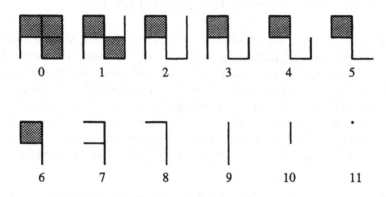

Fig. 7. Each of the complexes 1 – 11 is obtainable from its predecessor by an elementary collapse. Since the final complex in this sequence consists of a single 0-xel, each of the complexes 0 – 11 is collapsible.

Collapsing and collapsibility are of interest to us largely because of the following well known fact:

Proposition 10. *If a complex K collapses to a complex L, then there is a continuous deformation of $\bigcup K$ to $\bigcup L$ that never moves the points of $\bigcup L$.*

Proof (sketch). It is enough to show that such a continuous deformation exists whenever L is obtainable from K by a single elementary collapse. Suppose L is obtained from K by an elementary collapse at a free face f of an element $q \in K$. Let d be the dimension of q and let $f_0, f_1, \ldots, f_k$ be an enumeration of the $(d-1)$-dimensional faces of q, where f_0 is the free face f. For $0 \le i \le k$ let P_i be the affine hull of f_i, and let H_i be the closed d-dimensional half-space in the affine hull of q such that H_i is bounded by the hyperplane P_i and $q \subseteq H_i$. It can be shown that the half-spaces H_i exist, and that $q = \bigcap_{i=0}^{i=k} H_i$.

Let b be some point in $\bigcap_{i=1}^{i=k}(H_i - P_i) - H_0$. For each point x in q, it is readily confirmed that the straight line through b and x meets $\bigcup(\text{Boundary}(q) - \{f\})$ at just one point; let $g(x)$ be that point. For each point x in $\bigcup K - q$, let $g(x) = x$. It is readily confirmed that this defines a continuous map $g : \bigcup K \to \bigcup L$. For each x in $\bigcup K$ and each $t \in [0, 1]$, let $H(x, t) = tg(x) + (1 - t)x$. It is evident that H is continuous, that $H(x,0) = x$ and $H(x,1) \in \bigcup L$ for all x in $\bigcup K$, and that if $x \in \bigcup L$ then $H(x,t) = x$ for all t in $[0,1]$. Thus H defines a "continuous deformation of $\bigcup K$ to $\bigcup L$ that never moves the points of $\bigcup L$". □

7.2 A Sufficient Condition for Simpleness.

We will deduce the "if" part of the Main Theorem from the following sufficient condition for a xel to be simple. Note that in the following proposition n is not restricted to being ≤ 4.

Proposition 11. *Let n be any positive integer and let q be an n-xel in an n-image I such that* $\mathrm{Attach}(q, I)$ *is collapsible. Then q is simple in I.*

Our proof of this proposition depends on two lemmas.

Lemma 12. *Let c be the center of an n-xel q. Then for every xel complex $K \subseteq$* $\mathrm{Boundary}(q)$, $c * \mathrm{Boundary}(q)$ *collapses to $c * K$.*

Proof. If x is a xel in $\mathrm{Boundary}(q)$ that is not in K, then each xel in $\mathrm{Boundary}(q)$ that x is a face of is not in K. Starting with $c * \mathrm{Boundary}(q)$, we can perform elementary collapses at the $(n-1)$-xels in $\mathrm{Boundary}(q)$ that are not in K, then at the $(n-2)$-xels in $\mathrm{Boundary}(q)$ that are not in K, then at the $(n-3)$-xels in $\mathrm{Boundary}(q)$ that are not in K, and so on until elementary collapses have been done at all the xels in $\mathrm{Boundary}(q)$ that are not in K. We are then left with $c * K$, as required. □

Lemma 13. *Let c be the center of an n-xel q, and let K be a collapsible subcomplex of* $\mathrm{Boundary}(q)$. *Then the complex $c * K$ collapses to K.*

Proof. As K is collapsible, there is a sequence of elementary collapses that transforms K to a complex consisting of a single 0-xel p. Let m be the number of elementary collapses in this sequence, and let $K = K_0, K_1, \ldots, K_m = \{p\}$ be the sequence of complexes we go through. Let $f_0, f_1, \ldots, f_{m-1}$ be the free faces at which the elementary collapses occur (so that f_i is a free face of K_i). Then we can collapse $c * K$ to K using the sequence of $m+1$ elementary collapses at the free faces $c * f_0, c * f_1, \ldots, c * f_{m-1}, c$. □

Proof of Proposition 11. On putting $K = \mathrm{Attach}(q, I)$ in Lemmas 12 and 13 we deduce that:

(A) $c * \mathrm{Boundary}(q)$ collapses to $c * \mathrm{Attach}(q, I)$.
(B) $c * \mathrm{Attach}(q, I)$ collapses to $\mathrm{Attach}(q, I)$ (since $\mathrm{Attach}(q, I)$ is collapsible).

It follows from (A) and (B) that $c * \mathrm{Boundary}(q)$ collapses to $\mathrm{Attach}(q, I)$. Hence (by Proposition 10) there is a continuous deformation of $q = \bigcup(c * \mathrm{Boundary}(q))$ to $\bigcup \mathrm{Attach}(q, I)$, and q is simple. □

7.3 Completion of the Proof of the Main Theorem.

Let q be an n-xel in an n-image I, where $n \leq 4$, such that conditions (a – c) of the Main Theorem are satisfied. In view of Proposition 11, we can complete the proof of the Main Theorem by showing that Attach(q, I) must be collapsible.

The result is trivially valid when $n = 1$. When $n = 2$ the only possibilities for Attach(q, I) are those shown in Figure 8. In each case Attach(q, I) is evidently collapsible. So the Main Theorem is true for $n = 2$.

No edge   1 edge   2 edges   3 edges

Fig. 8. Possible forms of Attach(q, I) in the case where q is a pixel that satisfies conditions (a – c) of the Main Theorem.

Now suppose $n = 3$ or 4. Since q satisfies conditions (a – c) of the Main Theorem, the following conditions hold when $P =$ Attach(q, I):

    (A) $\bigcup P$ is nonempty and connected.
    (B) $\bigcup$ Boundary$(q) - \bigcup P$ is nonempty and connected.
    (C) $\bigcup P$ is simply connected.

We now show that any xel complex $P \subseteq$ Boundary(q) which satisfies (A), (B) and (C) is collapsible. Indeed, suppose the complex P is a minimal counterexample to this assertion. Then P satisfies (A – C) but is not collapsible. (B) implies that $P \neq$ Boundary(q). Since elementary collapses preserve properties (A – C), and P is a minimal counterexample, P cannot have any free face.

The Case $n = 3$. In this case it is evident (e.g., from Figure 2) that P cannot contain any 2-xel, for otherwise one of the 2-xels in P would have a free face in P. If P contains a 1-xel, then since each endpoint of a 1-xel in P must be an endpoint of another 1-xel in P (otherwise it would be a free face, contradiction) the 1-xels in P contain a cycle. But then $\bigcup P$ is not simply connected, which contradicts (C). Thus P contains only 0-xels. Since $\bigcup P$ is nonempty and connected (by (A)), P consists of exactly one 0-xel. But then P is (trivially) collapsible after all, which is a contradiction. Thus the minimal counterexample P cannot exist, and so the Main Theorem holds when $n = 3$.

The Case $n = 4$. In this case we see from Figure 3 that the minimal counterexample P cannot contain any 3-xel, for otherwise one of the 3-xels in P would have a free face in P. If P also contains no 2-xel, then we can use the same argument as we just used in the case $n = 3$ to derive a contradiction.

Now we suppose P contains a 2-xel, and deduce another contradiction. Let Schlegel(P) be a Schlegel diagram of P.

It is convenient to refer to 1-xels as *edges* and 0-xels as *vertices*. Each edge in Boundary(q) is incident on just three 2-xels of Boundary(q), and hence on zero, one, two or three 2-xels in P. No edge is incident on just one 2-xel in P (otherwise the edge would be a free face of that 2-xel). We claim that no edge is incident on three 2-xels in P.

Assume for the moment that we can prove this. Then each edge is incident on no 2-xel in P or just two 2-xels in P. It follows from a standard "Jordan surface" theorem—the Alexander Duality Theorem (see, e.g., [2])—that the set $\mathbf{R}^3 - \bigcup$ Schlegel(P) is disconnected, so that $\bigcup$ Boundary(q) $-\bigcup P$ is disconnected, contrary to (B).

It remains only to justify our claim that no edge is incident on three 2-xels in P. Call an edge a *3-edge* of P if it is incident on three 2-xels in P, and call the corresponding element of Schlegel(P) a *3-edge* of Schlegel(P). (Note that all 2-xels of Boundary(q) that are incident on a 3-edge of P are in P.) We want to show that no 3-edges of P exist. We first prove:

Lemma 14. *Let y be any 3-xel in* Boundary(q). *Then at most four of the six 2-xels that are faces of y can be in P.*

An immediate corollary of this lemma is:

Corollary 15. *In each of (a, b, c) below, the four thick black edges cannot all be 3-edges of* Schlegel(P)*:*

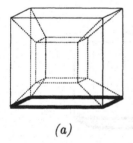 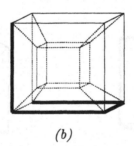 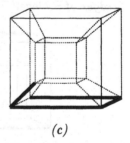

(a)   (b)   (c)

Proof of Lemma 14. It will probably be helpful to refer to Figure 3 in following this proof.

Since we know P contains no 3-xel, there must be a 2-d face f_1 of y that is not in P, for otherwise $\bigcup$ Boundary(q) $-\bigcup P$ would not be connected. Suppose the other five 2-d faces of y are all in P. Let y_1 be the 3-xel in Boundary(q) that shares the face f_1 with y, let f_2 be the 2-d face of y_1 opposite f_1, let y_2 be the 3-xel in Boundary(q) that shares the face f_2 with y_1, let f_3 be the 2-d face of y_2 opposite f_2, let y_3 be the 3-xel in Boundary(q) that shares the face f_3 with y_2, and let f_4 be the 2-d face of y_3 opposite f_3. (Thus f_4 is the 2-d face of y opposite f_1.)

Since the four 2-d faces of y that are adjacent to f_1 are all in P but none of them has a free face in P, and since $f_1 \notin P$, the four 2-d faces of y_1 that are adjacent to f_1 and f_2 are all in P. Hence $f_2 \notin P$, for otherwise $\bigcup \text{Boundary}(q) - \bigcup P$ would not be connected. So since none of the four 2-d faces of y_1 that are adjacent to f_2 has a free face in P, the four 2-d faces of y_2 that are adjacent to f_2 and f_3 are all in P. Hence $f_3 \notin P$ (as $\bigcup \text{Boundary}(q) - \bigcup P$ is connected) and the four 2-d faces of y_3 that are adjacent to f_3 and f_4 are all in P (as P has no free face). But this implies that $\bigcup P$ separates the interiors of the 3-xels y, y_1, y_2 and y_3 from the interiors of the other four 3-xels of $\text{Boundary}(q)$, contrary to the hypothesis that $\bigcup \text{Boundary}(q) - \bigcup P$ is connected. So at most four of the 2-d faces of y can be in P. □

Our next step in proving that no 3-edges of P exist is:

Lemma 16. *No vertex of* $\text{Boundary}(q)$ *is incident on just one 3-edge of P.*

Proof. Let e_1, e_2, e_3 and e_4 be the four edges of $\text{Boundary}(q)$ that are incident on some vertex v. We will prove the lemma by showing that the number of e's which are 3-edges of P is even. For $1 \leq i \leq 4$, let $f(e_i)$ be the number of 2-xels in P that have e_i as an edge (so that $f(e_i) = 0, 1, 2$ or 3). Now $f(e_i) \neq 1$ (otherwise e_i would be a free face of P), so $f(e_i)$ is odd if and only if e_i is a 3-edge of P.

Note that $f(e_1) + f(e_2) + f(e_3) + f(e_4)$ is even since each 2-xel incident on v is incident on exactly two of the e's and is therefore counted just twice in this sum, and 2-xels not incident on v are not incident on any e and are not counted. Hence the number of $f(e_i)$'s that are odd is even, and so the number of e_i's that are 3-edges of P is even. □

Corollary 17. *In each of (a) and (b) below, the thick black edges cannot all be 3-edges of* $\text{Schlegel}(P)$:

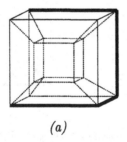 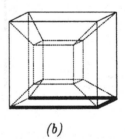

(a)   (b)

Proof. The thick black edges in (a) cannot all be 3-edges of $\text{Schlegel}(P)$ by Lemma 14. If the thick black edges in (b) were all 3-edges of $\text{Schlegel}(P)$ then by Lemma 16 all four thick black edges in one of the cases of Corollary 15 would be 3-edges of $\text{Schlegel}(P)$, contrary to that result. □

Now we suppose P has a 3-edge and deduce a contradiction. By Lemma 16, Corollary 17 and symmetry, we may assume the three light gray edges in the left-hand diagram of Figure 9 are 3-edges of $\text{Schlegel}(P)$. Then it follows, again from Lemma 16 and Corollary 17, that the five thick black edges in that diagram are also 3-edges of $\text{Schlegel}(P)$. Hence the 2-cell a in the right-hand diagram of Figure 9 is in $\text{Schlegel}(P)$ (since a is incident on a 3-edge of $\text{Schlegel}(P)$).

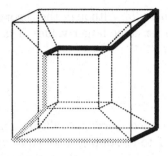 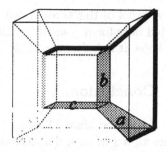

Fig. 9. If P has a 3-edge then we may assume the three gray edges in these diagrams are 3-edges of Schlegel(P). This implies that the five thick black edges are also 3-edges of Schlegel(P), and hence that a and either b or c in the right-hand diagram are in Schlegel(P). But this leads to a contradiction of Lemma 14.

As a is in Schlegel(P) and P has no free faces, at least one of the 2-cells b and c is also in Schlegel(P). Four of the six 2-cells that are faces of the inner 3-cell in these diagrams are incident on a 3-edge of Schlegel(P), and are therefore in Schlegel(P). Since either b or c must also be in Schlegel(P), we see that at least five of the six 2-cells that are faces of the inner 3-cell are in Schlegel(P). This contradiction of Lemma 14 proves that there are no 3-edges of P and completes the proof of the Main Theorem in the case $n = 4$.

8 Another Discrete Characterization of Simple Xels.

A. W. Roscoe and the author have discovered another discrete characterization of simple 1's in n-dimensional binary images, for $n \leq 4$. To state this characterization, define a *generalized image* to be a set I of xels such that if x is any xel in I then no face of x is in I. (Note that an n-dimensional binary image is a generalized image, for any positive integer n. A xel complex is not a generalized image unless it consists only of 0-xels, but the set of maximal xels of any xel complex is a generalized image.)

If q is any xel in a generalized image I, then the *attachment image* of q in I, denoted by ati(q, I), is defined to be the generalized image given by the set of all the maximal xels in Boundary(q) $\cap \bigcup$\{Boundary(x) $\mid x \in I - \{q\}$\}.

Definition 18. A xel q in a generalized image I is *inductively simple* in I if ati(q, I) $\neq \emptyset$ and the xels in ati(q, I) can be arranged in a sequence $a_0, a_1, \ldots, a_k$ such that, for $0 \leq i \leq k - 1$, a_i is inductively simple in the generalized image ati(q, I) $- \{a_j \mid 0 \leq j < i\}$. (Note that this defining condition is vacuously true if ati(q, I) consists of just one xel; this is the base case of the inductive definition.)

Theorem 19 (Kong and Roscoe). *For $n \leq 4$, a 1 in an n-dimensional binary image I is simple in I if and only if it is inductively simple in I.*

A proof of this theorem will be given elsewhere [5]. Although the theorem is stated only for $n \leq 4$, the result may well hold for some larger values of n. But we know it fails for $n \geq 24$ [5].

9 Conclusion.

For $n \leq 4$, the Main Theorem (Theorem 7) gives necessary and sufficient conditions for a 1 in an n-dimensional binary image I to be simple; if we know which of the $3^n - 1$ neighbors of a given 1 of I are also 1's of I, then it is easy to use these conditions to determine, by visual inspection, whether the given 1 is simple. An alternate version of the Main Theorem (Theorem 9) gives a practical algorithm for solving the 4-dimensional case of this problem.

A different discrete characterization of simple 1's in binary images of dimension ≤ 4, due to the author and A. W. Roscoe, has also been stated.

10 Acknowledgment.

The 4-dimensional case of the Main Theorem was conjectured to the author by A. W. Roscoe in 1995. The author is pleased to acknowledge stimulating discussions with Prof. Roscoe, in person and by e-mail, on this topic.

References

1. M. A. Armstrong, *Basic Topology*, Springer-Verlag, New York, 1983.
2. S. S. Cairns, *Introductory Topology*, Ronald Press, New York, 1968.
3. T. Y. Kong, On topology preservation in 2-D and 3-D thinning, *International Journal of Pattern Recognition and Artificial Intelligence* **9**, 1995, 813 – 844.
4. T. Y. Kong and A. W. Roscoe, Characterizations of simply-connected finite polyhedra in 3-space, *Bulletin of the London Mathematical Society* **17**, 1985, 575 – 578.
5. T. Y. Kong and A. W. Roscoe, Simple Points in 4-Dimensional (and Higher-Dimensional) Binary Images. Paper in preparation.
6. G. Malandain and G. Bertrand, Fast characterization of 3D simple points, *Proceedings, 11th IAPR International Conference on Pattern Recognition*, Volume III, The Hague, The Netherlands, 1992, 232 – 235.
7. A. Rosenfeld, Connectivity in digital pictures, *J. ACM* **17**, 1970, 146 – 160.
8. P. K. Saha, B. Chanda and D. D. Majumder, *Principles and Algorithms for 2D and 3D Shrinking*, Technical Report TR/KBCS/2/91, N. C. K. B. C. S. Library, Indian Statistical Institute, Calcutta, India, 1991.
9. Y. F. Tsao and K. S. Fu, A 3D parallel skeletonwise thinning algorithm, *Proceedings, IEEE Computer Society Conference on Pattern Recognition and Image Processing*, 1982, 678 – 683.

On the Reconstruction of Finite Lattice Sets from Their X-Rays

Peter Gritzmann*

Zentrum Mathematik, Technische Universität München,
D-80290 München, Germany
gritzman@mathematik.tu-muenchen.de

Abstract. We study various theoretical and algorithmic aspects of inverse problems in discrete tomography that are motivated by demands from material sciences for the reconstruction of crystalline structures from images produced by quantitative high resolution transmission electron microscopy.

In particular, we discuss questions related to the ill-posedness of the problem, determine the computational complexity of the basic underlying tasks and indicate algorithmic approaches in the presence of NP-hardness.

1   Introduction

While having a number of interesting applications in and connections to areas like general image processing, graph theory, scheduling, statistical data security, game theoryetc. (see e.g. [16], [28], [10], [21], [14], [15]), the main motivation for the present paper's discussion of the problem of reconstructing finite lattice sets from certain of their marginal sums is the demand from material sciences to reconstruct crystalline structures given through their images under high resolution transmission electron microscopy in a certain limited number of directions. In fact, [29] and [22] show how a quantitative analysis of images from high resolution transmission electron microscopy can be used to determine the number of atoms on atomic columns in certain directions. The goal is to use this technique for quality control in VLSI-technology. In particular, the interfacial topography of a material is vital in the manufacture of silicon chips.

So, in principle, we are given the information how many atoms there are on each line parallel to a certain direction for a certain small number of different directions. To be more precise, let $n \in \mathbb{N}$, $n \geq 2$, let F be a finite subset of $\mathbb{Z}^n$, let S be a line through the origin, and let $\mathcal{A}(S)$ denote the set of all lines of Euclidean n-space $\mathbb{E}^n$ that are parallel to S. Then the *(discrete) X-ray of F parallel to S* is the function $X_S F : \mathcal{A}(S) \to \mathbb{N}_0 = \mathbb{N} \cup \{0\}$ defined by

$$X_S F(T) = |F \cap T| = \sum_{x \in T} \mathbf{1}_F(x),$$

* Supported in part by a Max Planck Research Award and by the German Federal Ministry of Education, Science, Research and Technology Grant 03-GR7TM1.

for $T \in \mathcal{A}(S)$. Of course, in practice only the cases $n = 2$ and $n = 3$ are relevant.

Figure 1 shows a 3D-sample that has been reconstructed from three of its X-rays by methods outlined in Section 5.

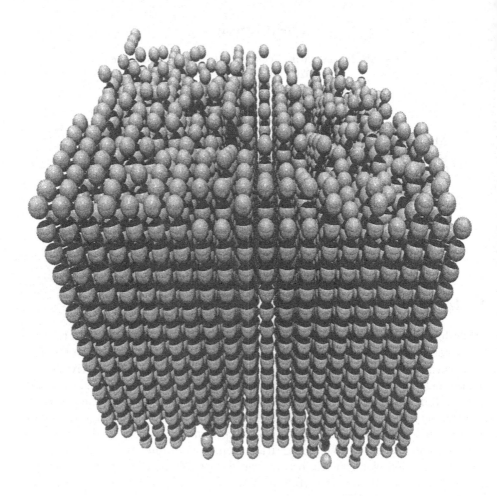

Fig. 1. *A 3D-sample*

It should be mentioned that the restriction to the lattice $\mathbb{Z}^n$ is not a crucial one. One the one hand, the problem is affinely invariant and, on the other hand, the whole model can be rephrased in a purely combinatorial form that relies only on incidences anyway.

Clearly, the best known and most important part of the general area of tomography is *computerized tomography*, an invaluable tool in medical diagnosis and many other areas including biology, chemistry and material sciences. While the mathematics of computerized tomography is quite well understood and utilized in every day practice, bringing down the resolution to the atomic scale changes this inverse problem drastically. Continuous methods do not seem appropriate anymore. In the following we will show, how the discreteness effects the problem.

2 Uniqueness Results

Clearly, when real world data is given, it is not enough to check consistency (up to data errors) and reconstruct an object compatible with the given X-ray data. It is absolutely necessary to have some uniqueness and stability results as well. It would not be a great help to be able to reconstruct an object that satisfies all the given X-ray constraints and fits the data perfectly if there were many other solutions of very different shape and nature. So let us begin by dealing with the question of unique determination of finite lattice sets by their X-rays.

In the following let $\mathcal{F}^n = \{F : F \subset \mathbb{Z}^n \wedge F \text{ is finite}\}$ and $\mathcal{S}^n = \{\text{lin}\{u\} : u \in \mathbb{E}^n \setminus \{0\}\}$. The elements of $\mathcal{F}^n$ are called *lattice sets*. Quite typically, we have some additional a priori information available. This is modeled by considering a suitable subset $\mathcal{G}$ of $\mathcal{F}^n$. For instance, $\mathcal{G}$ may incorporate some contiguity condition that reflects that the crystalline structures that are to be reconstructed do not consist of 'scattered' atoms but are highly connected. In most cases it will also be necessary to consider subsets $\mathcal{T}$ of $\mathcal{S}^n$ since electron microscopic images of high enough resolution can only be obtained in certain directions. An important such subset of $\mathcal{S}^n$ is the set $\mathcal{L}^n = \{\text{lin}\{u\} : u \in \mathbb{Z}^n \setminus \{0\}\}$ of *lattice lines*.

We say that $\mathcal{G}$ is *determined by m X-rays with respect to $\mathcal{T}$* if and only if there exist $S_1, \ldots, S_m \in \mathcal{T}$ such that the following holds: When $F_1, F_2 \in \mathcal{G}$ and $X_{S_j} F_1 = X_{S_j} F_2$ for $j = 1, \ldots, m$, then $F_1 = F_2$.

Let us begin with a trivial uniqueness results that indicates already the fundamental difference between discrete and continuous tomography.

2.1 *With respect to $\mathcal{S}^n$, the class $\mathcal{F}^n$ is determined by one X-ray.*

Of course, an X-ray line in a non-lattice direction either misses $\mathbb{Z}^n$ or, if it contains a lattice point, this lattice point is the unique lattice point on this line. From a certain point of view, (2.1) solves the problem. However, X-rays in non-lattice directions are not practical at all. The resolution coming from such X-rays would not be good enough, the image would typically be completely blurred. In fact good resolution can only be achieved in practice in certain main directions of the lattice.

Another quite simple uniqueness theorem is due to [26].

2.2 *For $n = 2$, any $|F| + 1$ X-rays in pairwise non-parallel directions determine F within $\mathcal{F}^n$.*

Some extensions of this result are contained in [5]. The problem with (2.2) in practice is that the typical atomic structures that have to be reconstructed comprise about 10^6 to 10^9 atoms. That means that we would need an extremely large number of X-ray images to be certain that the object is uniquely determined. However, after about 3 to 5 images taken by high resolution electron microscopy, the object is destroyed by the energy of the radiation, i.e. the object changes and after just a few X-rays it is no longer the original object that is 'seen' by the subsequent X-rays. In fact, it is easy to see that a fix number of X-rays is not sufficient to determine finite lattice sets uniquely.

In order to obtain positive results there seem to be only two options. The first option is to restrict the class of lattice sets under consideration. This general approach is quite reasonable since there is lots of information from physics waiting to be utilized. In terms of mathematical uniqueness results there is the following theorem of [12] for the class C^n of *convex* lattice sets, i.e. finite subsets F of $\mathbb{Z}^n$ such that $F = \mathbb{Z}^n \cap \text{conv}(F)$.

2.3 *The class C^n is determined by suitable 4 and any 7 X-rays in pairwise non-parallel coplanar lattice directions.*

It is an open problem where the later result persists when the assumption of coplanarity of the lattice directions is abandoned. In practice this restriction is satisfied since for technical reasons the 'tilting' of the microscope is confined to rotations about an axis, hence all X-ray directions lie in a common plane. This means, further, that the reconstruction can be done slice by slice, hence the underlying problem is essentially 2-dimensional. Let us point out that among the 'good' sets of directions in terms of (2.3) there are many that do have components of quite small absolute value which can be handled in practice. Examples for such sets in $\mathbb{E}^2$ are $\{(1,0),(1,1),(1,2),(1,5)\}$, $\{(1,0),(2,1),(0,1),(-1,2)\}$ and $\{(2,1),(3,2),(1,1),(2,3)\}$, see [12] and also [3]. While this result is quite reassuring, it is only practical in a very restricted setting. There may be some applications, for instance in colloid physics, but the main demand for mathematical methods for solving the inverse problems of discrete tomography comes from applications that involve the reconstruction of highly non convex objects. In particular, quality control in certain stages of chip production involves the detection of 'bumps' on the surface of silicon chips, hence convexity is not an appropriate condition in this situation.

It may be possible to weaken the assumptions of convexity so that only convexity in the X-ray directions together with strong connectivity assumptions are needed. At present, however, the above result may be mathematically satisfactory, in practice it is largely irrelevant.

So far, we have considered the problem of unique determination under the assumption, that we choose the directions in which X-rays are taken beforehand. A reasonable approach may, on the other hand, be to take the first X-ray in an arbitrary direction but then use the information gained from analyzing the image in order to determine the next direction in which an X-ray is taken. For the third direction, one could then use the complete information given by the first two X-rays and so on. This approach of *successive determination* leads to strong uniqueness results even for higher dimensional X-rays and even for more general sets then lattice sets, see [12]. In particular

2.4 $\mathcal{F}^n$ *can be successively determined by 2 X-rays.*

Again, while seeming to be satisfactory, this result is not practical at all. The reason is that the second X-ray has to be so 'skew' that one cannot produce images of high enough resolution in this direction.

Except for some results in more restricted situations, these are the only uniqueness results available at present. So what can be done?

Of course, the first question would be whether we have used all the crystallographic knowledge that there is. While this is certainly not the case is has not been possibly yet to identify reasonable mathematical constraints based on this crystallographic knowledge that reduce the relevant classes of lattice sets enough so as to lead to uniqueness theorems. In fact, there are examples showing that no fixed number of lattice X-rays suffices for determination that look like a solid crystalline blocks with 'just a few impurities.'

In view of the lack of satisfactory uniqueness theorems it seems that at least for the time being we have to settle for less. We may be satisfied by determining the 'core' of all solutions, the set of all *invariant points* that must belong to all solutions or at least to most in a sense that has to be made precise. We might also be satisfied by determining a 'typical' solution. We will come back to these aspects later. Another practically quite satisfying option could be to check uniqueness algorithmically. While we do not have general a priori uniqueness guarantees it is certainly true that in many practical applications sets to be reconstructed are determined uniquely by the available information. So, an efficient procedure to check uniqueness algorithmically might actually be all that is needed in practice. This brings up the algorithmic aspect of discrete tomography in the context of uniqueness. But it is certainly clear that efficient procedures for reconstruction are needed anyway. We will study algorithmic questions in the next sections.

3 Computational Complexity

In this section we state results dealing with the computational complexity of the questions of checking consistency of X-ray data, of determining uniqueness of given solutions and, of course, of finally reconstructing the objects. In the following we will focus on the full family $\mathcal{F}^n$, and the results will only be stated for that case. However, most of the results hold for a great variety of other subclasses $\mathcal{G}$ as well, without any significant change, see [14].

Suppose that $S_1, \ldots, S_m \in \mathcal{L}^n$ are $m \geq 2$ lines specified beforehand. In the inverse problem RECONSTRUCTION$(S_1, \ldots, S_m)$, we are given *candidate* functions

$$f_i : \mathcal{A}(S_i) \to \mathbb{N}_0, \qquad i = 1, \ldots, m$$

with finite support and want to find a set $F \subset \mathbb{Z}^n$ with corresponding X-rays or decide that no such F exists. Since for the purpose of computational complexity theory decision problems are more appropriate than reconstruction problems, we consider also the problem CONSISTENCY$(S_1, \ldots, S_m)$ whose instances are just the same but whose task is restricted to the decision whether a solution exists (without being obliged to produce one). Similarly, UNIQUENESS($S_1, \ldots, S_m$) asks whether, given a solution F, there exists another one. (Asking the question in this way puts the problem into the class $\mathbb{P}$.) Clearly, when investigating the computational complexity of such problems in the usual *binary Turing machine model* one has to describe suitable finite data structures and specify the problems accordingly. We refrain from doing this here but refer the reader to [14].

Here is a tractability results.

3.1 *Whenever $S_1, S_2 \in \mathcal{L}^n$, all three problems, CONSISTENCY(S_1, S_2), UNIQUENESS(S_1, S_2) and RECONSTRUCTION(S_1, S_2) can be solved in polynomial time.*

Proofs for the planar case can be found in [7], [16], [27], [28] or [2]. A particularly interesting way of looking at it is to interpret the problem as 2-PARTITION-MATROID-INTERSECTION. While this interpretation might not lead to the most efficient algorithm, it provides an elegant way of generalizing the results to higher dimensional X-rays and arbitrary dimensions. In fact, arbitrary linear optimization problems over the intersection of two matroids can be solved in polynomial time, see [8], [9]. So the case of just two X-rays is algorithmically tractable, see [13].

However, two X-rays usually do not determine finite lattice sets uniquely; hence, we would like to extend these tractability results to a greater number of X-rays. Unfortunately, there is a drastic jump in complexity from $m = 2$ to $m = 3$.

3.2 *For $n \geq 2$ and $m \geq 3$ different lines $S_1, \ldots, S_m$ in $\mathcal{L}^n$, the problems CONSISTENCY($S_1, \ldots, S_m$) and UNIQUENESS($S_1, \ldots, S_m$) are $\mathbb{NP}$-complete in the strong sense; RECONSTRUCTION($S_1, \ldots, S_m$) is $\mathbb{NP}$-hard.*

This intractability result due to [14] generalizes and sharpens previously known results. N. Young (private communication) showed the $\mathbb{NP}$-hardness of CONSISTENCY(S_1, S_2, S_3, S_4) when $S_1, \ldots, S_4$ are the four coordinate axes in $\mathbb{E}^4$. He used this result to obtain the $\mathbb{NP}$-hardness of a consistency problem for $n = 2$ and $m = 4$; however, the fourth direction is part of the input, so this is much weaker than the corresponding case of (3.2). Also, it is shown in [21, Section 4.1] (in the context of contingency tables), by a transformation from LATIN-SQUARE, that CONSISTENCY(S_1, S_2, S_3) is $\mathbb{NP}$-complete when S_1, S_2, S_3 are the coordinate axes in $\mathbb{E}^3$.

The hardness results of [14] are given by transformations from the well known $\mathbb{NP}$-complete problem 1-IN-3-SAT. The constructions shows that hardness does occur in situations which are not too far off from practice. In fact, the constructed objects are again solid crystals with just a few impurities and can represent, in principle, physically reasonable objects. Therefore (3.2) seems to be an appropriate explanation of the algorithmic difficulties observed in practice. An extension of these hardness results is given in [19].

There are some other related complexity results, particularly those of [4] and [31] for polyominoes in the plane.

4 Algorithmic Approaches in the Presence of NP-Hardness

Even though our problems are NP-hard, they have to be solved in practice. Various approaches have been suggested for dealing with CONSISTENCY$(S_1, \ldots, S_m)$, UNIQUENESS$(S_1, \ldots, S_m)$ and RECONSTRUCTION$(S_1, \ldots, S_m)$ in practice. See [19] for an account of the success and failure of many of those techniques. For simplicity, we restrict the following exposition to the problem of checking data consistency, and we will always assume that $S_1, \ldots, S_m$ are $m \geq 2$ different lattice lines.

It is easy to see, that CONSISTENCY$(S_1, \ldots, S_m)$ can be formulated as an integer linear programming problem whose variables correspond to the possible positions of elements of a solution. In fact, the *grid G* associated with a given instance of the problem consists of all (finitely many) lattice points that arise as points of intersection of m lines parallel to $S_1, \ldots, S_m$, respectively, whose candidate function value is nonzero, i.e.

$$G = \mathbb{Z}^n \cap \bigcap_{i=1}^{m} \bigcup_{T \in \mathcal{T}_i} T,$$

where $\mathcal{T}_1, \ldots, \mathcal{T}_m$ denote the supports of the given candidate functions $f_1, \ldots, f_m$, respectively. The incidences of G and $\mathcal{T}_i$ can be encoded by an incidence matrix A_i. If G consist of, say, N points, $M_i = |\mathcal{T}_i|$ for $i = 1, \ldots, m$, and $M = M_1 + \ldots + M_m$, then the incidence matrices A_i are in $\{0,1\}^{M_i \times N}$, and can be joined together to form a matrix $A \in \{0,1\}^{M \times N}$. Identifying a subset of G with its characteristic vector $x \in \{0,1\}^N$, the reconstruction problem amounts to solving the integer linear feasibility program

$$Ax = b, \quad \text{s.t. } x \in \{0,1\}^N, \tag{1}$$

where $b^T = (b_1^T, \ldots, b_m^T)$ contains the corresponding values of the candidate functions $f_1, \ldots, f_m$ as the right hand sides of $A_1, \ldots, A_m$, respectively.

Since linear programming problems can be solved in polynomial time the first natural approach is to consider the *LP-relaxation*

$$Ax = b \wedge 0 \leq x \leq 1, \tag{2}$$

of (1), where, as usual, '$\leq$' is to be understood componentwise; see [11]. Since linear programming codes are available for solving these problems very efficiently for all sizes of crystalline structures that are relevant in practice, computation time is not much of an issue for this heuristic. However, the solution is usually far from being integer and it does not seem completely justified to interpret fractional components of solutions as probabilities that the corresponding points belong to a (typical) solution.

In an improvement strategy N. Young studied the effect of such an interpretation followed by a subsequent randomized rounding. Here an atom is placed at the corresponding lattice point with the probability coming from the fractional solution produced by the LP-solver. This way an approximative solution to the consistency problem is produced. Compared to other heuristics for the problems in discrete tomography, known bounds for such solutions are, however, in general rather weak. We will give some bound on approximation errors for various heuristics in Section 5.

Of course, one cannot expect too tight a priori error bounds for polynomial-time approximative heuristics in general. But again, for all practical purposes it is nice but not absolutely necessary to have such a priori bounds. In fact, a method that would, in the course of the run of the algorithm, provide upper and lower bounds might be all that is needed. One can then run the algorithm on the given data until the gap between upper and lower bounds is small enough and then terminate with an approximate solution including a performance guaranty. The next section will give the basic idea of such an algorithm, the branch-and-cut method.

5   Polytopes in Discrete Tomography

The idea behind the *branch-and-cut* method is to try to approximate the polyhedral structure of the convex hull

$$P(A, b) = conv\{x \in \{0, 1\}^N : Ax = b\},$$

of all solutions of (1). In the algorithm, we begin with an LP-relaxation of the integer programming problem (1), compute a solution, and check whether it is already in $\{0, 1\}^N$. If this is the case, we stop with a consistent solution for the given instance. Otherwise we try to find a (facet-defining) constraint for $P(A, b)$ that is violated by the LP solution. This means, we try to find a cutting plane that is in a sense best possible for providing a deep cut. If such a *separating hyperplane* can be found, we add the corresponding constraint to the current LP and repeat the procedure. It may actually happen that the LP-solution found is not in $\{0, 1\}^N$ but we are still not able to produce a separating hyperplane. In that case, we use a branch-and-bound paradigm and proceed in the same algorithmic framework as before – just splitting the problem into subproblems. Such an approach has been successfully utilized for many problems in combinatorial optimization, most notably for the *traveling salesman problem*, see [20], [25] for surveys on the polyhedral theory and on polyhedral computations for the traveling salesman problem, and see [6] for a geometric introduction into the general concept underlying polyhedral combinatorics.

Since, in general, computing the dimension of $P(A, b)$ is NP-hard by (3.2) and since it does not seem wise to try to 'jump to a solution' in one step anyway, $P(A, b)$ is replaced in the polyhedral study [17] by

$$T(A, b) = conv\{x \in \{0, 1\}^N : Ax \leq b\}.$$

Then, of course, we have to model that we are only interested in a solution of maximum cardinality in order to obtain an equivalent problem. This can be done by adding the objective to maximize the funktion $x \mapsto e^T x$, where $e = (1, \ldots, 1)^T \in \mathbb{E}^N$.

The *tomography polytopes* $T(A, b)$ are quite special. Clearly, all tomography polytopes are subpolytopes of the standard cube $\{0, 1\}^N$ whose coefficient matrix A is a 0-1-matrix with a particular structure. Specifically, all submatrixes corresponding to just two directions are *totally unimodular* – another explanation why the case of two directions is simple. It follows from [24] that the combinatorial *diameter* of a tomography polytope is at most N. This means that, in principle, an edge-path could be found leading from 0 to a solution of the problem that is rather short. The main problem of course is, that it is not known how to actually choose the pivot element at each step in order to find such short paths. Of course the underlying physics induces additional structure that can be utilized in the study of the polytopes.

Further, using all sorts of preprocessing techniques, the practically relevant dimensions of these polytopes can be reduced to about 10^4. Judged on the base of the sizes of successfully solved instances of the traveling salesman problem this seems encouragingly small. While the traveling salesman polytopes correspond to the complete graph on the given number of 'cities' and are hence universal, tomography polytopes depend, on the other hand, on the right-hand side b. So the most important goal of polytopal investigations is to find large systems of valid inequalities that are facet-defining under week conditions on the right-hand side. In [17] various classes of facet-defining inequalities are determined under very weak assumptions on b. Usually it is only necessary to require that all components of b are at least 2 or 3 and that no X-ray line is completely filled with atoms. Using these branch-and-cut techniques one can reconstruct moderately big crystalline structures already. The example of Figure 1 was produced by a (not yet fully developed) algorithm of this kind. It shows a 3D-phantom – whose X-ray data in directions $(0, 1, 0), (1, 4, 0)$ and $(-1, 4, 0)$ were given to us by P. Schwander – that turns out to be uniquely determined by the X-rays in these directions. (One should regard the front plane as the xy-plane while the different z-coordinates correspond to planes parallel to the xy-plane).

6 Approximation Algorithms with A Priori Performance Guarantees

In the following we study some simple heuristics that can be fully analyzed and give approximation guarantees that yield a constant relative error. It is clear that these techniques by themselves will not solve the problem since their performance in terms of approximating a solution is still rather poor. However, it is first of all reassuring that simple techniques yield such a small relative error, and on the other hand such techniques can be used as a preprocessing step in order to find good bounds in any branch-and-bound based approach.

The two problems that we are going to consider now are BEST-INNER-FIT$(S_1, \ldots, S_m)$ [BIF] and BEST-OUTER-FIT$(S_1, \ldots, S_m)$ [BOF]. Given candidate functions $f_1, \ldots, f_m$, the tasks of these problems are to find a set $F \subset G$, respectively. For [BIF], $|F|$ must be maximal such that $X_{S_i}F(T) \leq f_i(T)$ for all $T \in \mathcal{T}_i$ and $i = 1, \ldots, m$, while for [BOF], $|F|$ must be minimal such that $X_{S_i}F(T) \geq f_i(T)$ for all $T \in \mathcal{T}_i$ and $i = 1, \ldots, m$.

For an analysis of these and other approximation problems for discrete tomography see [18]. The first strategy for [BIF] that is fully analyzed is that of the *greedy* heuristic: place points into the grid of the given instance whenever that is possible without violating the constraints. Then one obtains the following bound.

6.1 *If F is a solution of [BIF] and L is a greedy solution then $m|L| \geq |F|$.*

It may be worthwhile to point out that the greedy strategy is very flexible and allows various specifications for breaking the ties between different choices for points to be placed next. For example, the X-ray data can be used in a way that is very similar to back-projection techniques to express preferences. In addition, connectivity of the solution (in a sense that is justified by the physical structure of the analyzed material) can be rewarded. Similarly, information of neighboring layers can be taken into account in a layer-wise reconstruction of a 3D-object.

While the above result follows from work of [23], the next bound is due to [18]. It uses the notion of an *(increasing) r-change*: r points of a feasible set $F \subset G$ are deleted and $r + 1$ points of $G \setminus F$ are inserted instead without destroying the feasibility. Clearly, r-changes are more powerful than just greedy insertion (which actually can be regarded as a 0-change strategy).

6.2 *Let $t \in \mathbb{N}$, let F be a solution of [BIF] and let $L \subset G$ such that no further r-change is possible then*

$$|L| \geq \left(\frac{2}{m} - \epsilon(t) \right) |F|, \quad \text{where } \epsilon(t) \to 0 \text{ for } t \to \infty.$$

Similar results can be obtained for [BOF] see [18]. Rather than specifying the algorithms precisely – they are based again on greedy strategies, (decreasing) r-changes and matching techniques – let us just state a result for [BOF] that corresponds to (6.2).

6.3 *For every $\epsilon > 0$ there is a polynomial-time algorithm that, given an instance of [BOF] produces a set L that is feasible and whose cardinality exceeds that of an optimal solution by a factor at most*

$$\frac{2}{3} + \frac{1}{2} + \ldots + \frac{1}{m} + \epsilon.$$

Another simple but useful observation is that if we compute a basic optimal solution of the LP-relaxation $N - M$ of the 0-1-constraints of (1) are already satisfied.

7 Dealing with Nonuniqueness

So far, we have mainly studied the question of checking data consistency. A problem of similar (or even greater) importance, is that of dealing with nonuniqueness. Again, if the data given in practice allow many solutions that are extremely different, the relevance of finding one of them is not apparent any more. There are various suggestions for how to deal with the problem of nonuniqueness.

If we use the fact, that in the practical application for material sciences that motivated this study the directions in which the X-ray images are taken are coplanar, then the objects can be reconstructed layer by layer. Suppose that the first layer has been reconstructed and that it is uniquely determined by the given information (or known beforehand). Then it may be reasonable to assume, that the second layer does not vary too much from the first. We can easily model that by adding an objective function to our reconstruction problem for the second layer. In fact the incidence vector of the solution for the first layer provides a linear objective function whose maximization leads to a solution for the second layer that is in a certain sense closest to that for the first one. In case of just two X-rays such an approach was suggested by [16].

[11] uses an interior point method for solving the LP-relaxation

$$\max e^T x \text{ s.t. } Ax \leq b \wedge 0 \leq x \leq 1 \tag{3}$$

of the (monotone vorsion of the) reconstruction problem in order to identify positions that are uniquely determined by the given data. In fact such an approach produces a point that is interior to the optimal face of the LP-polytope. So if a component of the LP-solution is 0 or 1 then the same is true for all solutions and hence for all 0-1 solutions. While by [14] the problem of detecting whether a given subset of G belongs to all solutions is again NP-hard (see also [19]), it is demonstrated in [11] that for certain phantoms such an approach produces quite a large number of fixed variables. Note that the problem of deciding whether a subset of a possible solution can actually be extended to a full solution is again NP-complete, see [14]. It should be noted that the notion of *additivity* as introduced by [10] and extended by [1] is actually equivalent to 'substructure uniqueness' for the LP-relaxation.

All of the above mentioned techniques can be used and combined in practice to reduce nonuniqueness but it will certainly still be mandatory to take additional physical constraints into account in order to be able to produce solutions that resemble the actual physical objects.

8 Imprecise Data

Even if we manage to reduce nonuniqueness to a large extent and solve the reconstruction problem efficiently in practice, there is still the problem of imprecise data. Of course there is always some noise involved when physical measurements are taken but for the problem under consideration also some *systematic error* is present. In fact, the sample that is to be studied has to be prepared for electron microscopy by thinning it out so as to let the electron waves penetrate it. For technical reasons this is done in such a way that the object is much thinner in the middle than at the boundary where it is attached to some position control element. This means that an X-ray taken in some direction 'sees' the crystalline structure in the middle of the object but is absorbed towards the boundary. Taking a second direction, one has the same effect, just that the parts of absorption at the boundary do not coincide. That means that in practice the data that one gets from the measurements actually correspond to slightly different objects.

There are various ways of modeling data inconsistency and error. General techniques for dealing with ill-posed problems can be applied by using an objective function that minimizes the distance of the computed right-hand side for a found solution from the given right-hand side data. This can be done in various different ways. A promising suggestion of this kind is due to [30] since it allows to interpret solutions as best-approximations in the sense of maximum likelihood. The method is, however, again based on the LP-relaxation of the underlying 0-1-programming problem. It will be particularly intriguing to find out on the basis of real-world data to what extent 'severe artifacts' are introduced by methods based on the underlying LP-relaxation. It is certainly clear that LP-based techniques are generally not capable of capturing the discreteness of the problem. Maximum likelihood strategies on all possible 0-1 solutions are on the other hand algorithmically quite demanding.

9 Final Remarks

The problems of discrete tomography are practically important, mathematically rich (involving methods from areas like combinatorial optimization, number theory, geometry, combinatorics, stochastics, etc.) and algorithmically challenging. During the last few years there has been considerable progress in attacking these problems. However, there is still a lot of work to be done to finally create a tool that is as developed and satisfactory for the application in material sciences as is computer tomography in its medical and other applications.

References

1. R. Aharoni, G.T. Herman, and A. Kuba, *Binary vectors partially determined by linear equation systems*, Discrete Math. **171** (1997), 1–16.
2. R.P. Anstee, *The network flow approach for matrices with given row and column sums*, Discrete Math. **44** (1983), 125–138.
3. E. Barcucci, A. Del Lungo, M. Nivat, and R. Pinzani, *X-rays characterizing some classes of digital pictures*, preprint.
4. _____, *Reconstructing convex polyominoes from their horizontal and vertical projections*, Th. Comput. Sci. **155** (1996), 321–347.
5. G. Bianchi and M. Longinetti, *Reconstructing plane sets from projections*, Discrete Comput. Geom. **5** (1990), 223–242.
6. T. Burger and P. Gritzmann, *Polytopes in combinatorial optimization*, Geometry at Work (C. Gorini, ed.), Math. Assoc. Amer., 1997, in print.

7. S.K. Chang, *The reconstruction of binary patterns from their projections*, Comm. ACM **14** (1971), no. 1, 21–25.

8. J. Edmonds, *Maximum matching and a polydedron with 0,1–vertices*, J. Research National Bureau of Standards **69B** (1965), no. 1 and 2, 125–130.

9. _____, *Matroid intersection*, Ann. Discrete Math. **4** (1979), 39–49.

10. P.C. Fishburn, J.C. Lagarias, J.A. Reeds, and L.A. Shepp, *Sets uniquely determined by projections on axes. II. discrete case*, Discrete Math. **91** (1991), 149–159.

11. P.C. Fishburn, P. Schwander, L.A. Shepp, and J. Vanderbei, *The discrete Radon transform and its approximate inversion via linear programming*, Discrete Appl. Math. **75** (1997), 39–62.

12. R.J. Gardner and P. Gritzmann, *Discrete tomography: Determination of finite sets by X-rays*, Trans. Amer. Math. Soc **349** (1997), 2271–2295.

13. R.J. Gardner, P. Gritzmann, and D. Prangenberg, *On the reconstruction of binary images from their discrete Radon transforms*, Proc. Intern. Symp. Optical Science, Engineering, and Instrumentation (Bellingham, WA), SPIE, 1996, pp. 121–132.

14. _____, *Computational complexity issues in discrete tomography: On the reconstruction of finite lattice sets from their line X-rays*, submitted, 1997.

15. _____, *On the computational complexity of determining polyatomic structures by X-rays*, submitted, 1997.

16. J.J. Gerbrands and C.H. Slump, *A network flow approach to reconstruction of the left ventricle from two projections*, Computer Graphics and Image Processing **18** (1982), 18–36.

17. P. Gritzmann and S. de Vries, *Polytopes for discrete tomography*, in preparation.

18. P. Gritzmann, S. de Vries, and M. Wiegelmann, *On the approximate reconstruction of binary images from their discrete X-rays*, in preparation, 1997.

19. P. Gritzmann, D. Prangenberg, S. de Vries, and M. Wiegelmann, *Success and failure of certain reconstruction and uniqueness algorithms in discrete tomography*, submitted, 1997.

20. M. Grötschel and M. Padberg, *Polyhedral theory*, The Traveling Salesman Problem (E.L. Lawler, J.K. Lenstra, A.H.G. Rinnoy Kan, and D.B. Shmoys, eds.), Wiley-Interscience, Chichester, 1985, pp. 251–305.

21. R.W. Irving and M.R. Jerrum, *Three-dimensional statistical data security problems*, SIAM J. Comput. **23** (1994), 170–184.

22. C. Kisielowski, P. Schwander, F.H. Baumann, M. Seibt, Y. Kim, and A. Ourmazd, *An approach to quantitative high-resolution transmission electron microscopy of crystalline materials*, Ultramicroscopy **58** (1995), 131–155.

23. B. Korte and D. Hausmann, *An analysis of the greedy heuristic for independence systems*, Ann. Discrete Math. **2** (1978), 65–74.

24. D. Naddef, *The Hirsch conjecture is true for $(0,1)$-polytopes*, Math. Prog. **45** (1989), 109–110.

25. M. Padberg and M. Grötschel, *Polyhedral computations*, The Traveling Salesman Problem (E.L. Lawler, J.K. Lenstra, A.H.G. Rinnoy Kan, and D.B. Shmoys, eds.), Wiley-Interscience, Chichester, 1985, pp. 307–360.

26. A. Rényi, *On projections of probability distributions*, Acta Math. Acad. Sci. Hungar. **3** (1952), 131–142.

27. H.J. Ryser, *Combinatorial properties of matrices of zeros and ones*, Canad. J. Math. **9** (1957), 371–377.

28. _____, *Combinatorial mathematics*, Mathem. Assoc. Amer. and Quinn & Boden, Rahway, NJ, 1963.

29. P. Schwander, C. Kisielowski, M. Seibt, F.H. Baumann, Y. Kim, and A. Ourmazd, *Mapping projected potential, interfacial roughness, and composition in general crystalline solids by quantitative transmission electron microscopy*, Physical Review Letters **71** (1993), 4150–4153.
30. Y. Vardi and D. Lee, *The discrete Radon transform and its approximate inversion via the ELM algorithm*, preprint, 1997.
31. G.J. Woeginger, *The reconstruction of polyominoes from their orthogonal projections*, preprint.

Aspects in Topology-Based Geometric Modeling
Possible Tools for Discrete Geometry ?

Pascal Lienhardt

Université de Poitiers, IRCOM-SIC (UMR CNRS 6615)
SP2MI, Bvd. 3, Téléport 2
F-86960 Futuroscope Cedex, France

Abstract. Topology-based Geometric Modeling is concerned with modeling subdivisions of geometric spaces. Methods are close to that of combinatorial topology, but for different purposes. We discuss some of these methods, their interests and drawbacks for Geometric Modeling, mainly aspects we think that could be of possible interest for Discrete Geometry.

1 Introduction

Geometric objects handled in Geometric Modeling are often structured point sets, since it is important for many applications, not only to represent a geometric object, but also to distinguish between different parts of the object, according to properties which are relevant for the application (e.g. mechanical, photometric, geometric properties). Important consequences exist for interactive or procedural modelers; for instance, the structure of a modelled object has to be taken into account when constructing it (it can be used for controlling construction operations).

Topology-based Geometric Modeling deals with subdivisions of geometric spaces, i.e. partitions of these spaces into cells: vertices, edges, faces, volumes, etc. which define a sort of discretization of the geometric space. The fundamentals of many approaches are close to that of Combinatorial Topology [Ale] [SeTh] [FrPi], but for different purposes, and consist in distinguishing between topological information and embedding information. Topological information can be captured by combinatorial structures and controlled by combinatorial operations. So implementation of these structures (and related algorithms) can be done without loss of information and without loss of properties.

More precisely, a subdivision is represented using a combinatorial structure which describes its topology (mainly the cells and their boundary relations), and embedding information is associated with topological information in order to describe the location of the subdivision into its embedding space (usually E^2 or E^3). Different methods exist for defining embedding: linear embedding, free-form spaces, hierarchized embedding (a subdivision is embedded onto an other subdivision, in order to distinguish between different representation levels, for instance).

An important consequence of this distinction between topology and embedding is that it is necessary to maintain a strong consistency between these two complementary aspects, and this has to be taken into account when designing construction operations [Tak].

Nevertheless, a main interest of topology-based geometric modeling is the fact that the relations between combinatorial structures and subdivisions of geometric spaces have been extensively studied by mathematicians. So, a lot of knowledge can be directly used: for instance, many results in combinatorial topology can be used for computing properties, which are useful for controlling (constructions of) modeled objects. Such controls are essential for most applications. Moreover, it is possible to modify several geometric algorithms (e.g. for rendering, finite element method), which classically perform numerical computations, so that they will exploit topological information, and perform combinatorial computations, increasing thus their reliability and efficiency.

Different types of subdivisions have to be handled, according to the applications: general complexes or particular manifolds, subdivided into any cells or regular ones (e.g. simplices): cf. Figure 1. Since 25 years, numerous structures, operations, modelers have been conceived in topology-based geometric modeling: first, cellular subdivisions of 2-dimensional manifolds are studied for Boundary Representation of solids [AFF] [Bau] [Bra] [Män] [Wei1], but recent works deal with simplicial [FePa] [LaLi1] [PBCF] or cellular geometric complexes [CCM] [CrRe] [ElLi1] [GCP] [LuLu] [Mar] [MuHi] [RoOC] [Wei2] or manifolds [ArKo] [DoLa] [GuSt] [HC] [Lie1] [Spe].

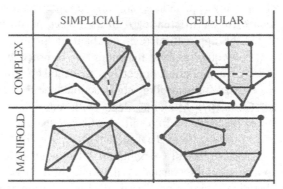

Fig. 1. Several types of topologically 2-dimensional subdivisions. Intuitively, complexes are any assemblies of cells (cells are regular, as simplices, or not). The neighborhood of any point of an n-dimensional manifolds is an n-ball (or half ball if the point lies on a boundary).

In order to illustrate previous remarks and assertions, we will focus on several works whose fundamentals are well-known in combinatorial topology, and we will discuss their interests for geomeric modeling. First, two combinatorial structures (simplicial sets, combinatorial maps) are presented : we discuss the definition of data structures, basic construction operations, and we show that classical geometric construction operations can be easily defined on these structures. Topological properties can be computed on these structures, and thus, geometric computations can be translated into combinatorial ones. Second, complexity is discussed: it is possible, using mechanisms based upon topological properties, to reduce the complexity of data structures and algorithms, without loss of information. Remarks are mentioned about implementation, programming and geometric modelers based upon the previous notions.

I think that topological aspects are common to Geometric Modeling and Discrete Geometry (as, for instance, very natural relations exist between free-form space modeling and topology-based modeling). Some recent works establish interesting links between these fields [Fio] [Fra] [Lac] [Bru] [Som]. In the following, some of them will be mentioned, but not carefully discussed, since I will focus on methods in Topology-based Geometric Modeling.

2 Subdivisions and combinatorial structures

In order to illustrate basic aspects in topology-based geometric modeling, we first present two combinatorial structures: semi-simplicial sets (resp. generalized combinatorial maps, or G-maps) are used for handling simplicial subdivisions of geometric complexes (resp. cellular subdivisions of manifolds).

2.1 Semi-simplicial sets

An n-dimensional semi-simplicial set [May] is a collection of sets $K = (K_k)_{k=0,...n}$ with maps (cf. Figure 2):
$$d_i: K_k \rightarrow K_{k-1}, k \geq 1, 0 \leq i \leq k,$$
satisfying the following relations:
$$d_i d_j = d_j d_{i-1}, j < i.$$

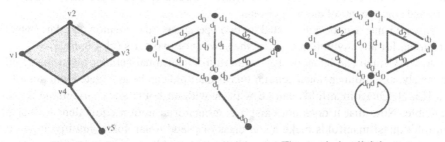

Fig. 2. 2-dimensional semi-simplicial set. The semi-simplicial set (middle) describes the topology of the triangulation (left). Each simplex correspond to a sequence of vertices; for instance, 2-simplices correspond to (v1,v2,v4) and (v2,v3,v4). $(v1,v2,v4)d_0 = (v2,v4)$, $(v1,v2,v4)d_1 = (v1,v4)$, $(v1,v2,v4)d_2 = (v1,v2)$. The semi-simplicial set (right) is constructed by identifying 0-simplices corresponding to v4 and v5. Local neighbourhood of a simplex corresponds to boundary and star notions. More generally, boundary operators and their inverses make it possible to explore more or less large neighbourhoods of parts of semi-simplicial sets.

Elements of K_k are k-dimensional abstract simplices (or k-simplices) ; maps d_i are face operators. A j-face of a k-simplex $(0 \leq j \leq k)$ is obtained by applying a sequence of (k-j) face operators. The face is principal if the sequence is empty, otherwise it is proper. The boundary of a simplex is the set of its proper faces: for instance, (k+1) face operators $(d_i)_{i=0,...k}$ associate the (k-1)-simplices of its boundary with every k-simplex. The star of a simplex is the set of all simplices of which it is a proper face.

Any semi-simplicial set can be constructed by two operations: the creation of a k-simplex (together with its boundary), and the identification of simplices.

The geometric realization of a semi-simplicial set is a topological space [Mil]. A topological cell, homeomorphic to the n-ball, is associated with any n-simplex. A CW-complex is defined by deducing identifications of cell boundaries from face relations in the semi-simplicial set.

2.2 Generalized maps

An n-dimensional generalized combinatorial map (or n-G-map: cf. [BrSi] [Lie2] [Vin]) G is a set D with bijections (cf. Figure 3):

$$\alpha_i: D \to D, \ 0 \leq i \leq n,$$

such that:

α_i is an involution on D, $0 \leq i \leq n$ (i.e. for any element d of D, $d\alpha_i\alpha_i = d$);

$\alpha_i\alpha_j$ is an involution, $0 \leq i < i+1 < j \leq n$.

D is a set of abstract objets called darts. A connected component of G incident to dart d of D is the set of all darts that can be reached by successively applying involutions α_i, starting from d. The i-cells of G are defined as the connected components of the (n-1)-G-maps $(D, \alpha_0, ..., \_\alpha_i, ..., \alpha_n)$, where $\_\alpha_i$ means that involution α_i is omitted in the sequence of involutions. If darts exist, such that they are invariant for α_i, G is with boundaries, else it is without boundaries. Boundaries of G are also defined as connected components of an (n-1)-G-map.

Any n-G-map can be constructed by two operations : the creation of a dart (which is invariant for any involution α_i), and the sewing of two darts (cf. Figure 3).

Any n-G-map can be associated with an n-dimensional cellular quasi-manifold; conversely, any n-dimensional cellular quasi-manifold can be associated with an n-G-map [Lie2]. Quasi-manifolds can be with or without boundaries, orientable or not orientable. Note that it does not exist a combinatorial notion equivalent to that of manifold: quasi-manifolds make a sub-class of pseudo-manifolds; intuitively, an n-dimensional cellular quasi-manifold can be constructed by sewing together n-cells by identifying (n-1)-cells, in such a way that at most two n-cells share an (n-1)-cell. In fact, we can prove that n-G-maps are equivalent to a sub-class of semi-simplicial sets, where simplices are structured into cells. So, notions and operations defined on semi-simplicial sets can be extended on G-maps. Moreover, other notions can be defined (e.g. boundary of a G-map), since G-maps make a sub-class of simplicial quasi-manifolds (i.e. semi-simplicial sets which can be constructed by creating n-simplices and by identifying (n-1)-simplices in such a way that at most two n-simplices share an (n-1)-simplex).

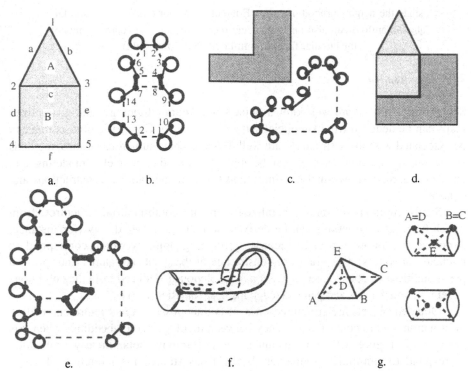

a.   b.   c.   d.

e.   f.   g.

Fig. 3. a: a cellular subdivision; b: the corresponding 2-G-map: a point corresponds to a dart, involution α_0 (resp. α_1,α_2) is symbolized by a dashed line (resp. line, thick line). All darts, except 4,5,7 and 8, are their own images by α_2. Intuitively, dart 5 (resp. $4=5\alpha_0,6=5\alpha_1,7=5\alpha_2$) corresponds to the 3-tuple (2,c,A) (resp. (3,c,A),(2,a,A),(2,c,B): cf. [Bri]). The 2-G-map is composed by a unique connected component. Vertex 2 (resp. edge c, face A) corresponds to {6,5,7,14} (resp. {4,5,7,8}, {1,2,3,4,5,6}), which darts are connected by α_1 and α_2 (resp. α_0 and α_2,α_0 and α_1). The 2-G-map, and its corresponding subdivision, has one boundary, defined by {1,2,3,9,10,11,12,13,14,6}. Involutions α_i make it possible to explore the neighbourhood of any dart and cell (and also to compute properties as duality, orientability, etc.: cf. below). c: a face and its corresponding 2-G-map. d and e: identification of vertices and edges, corresponding to sewing by α_2 (contrary to semi-simplicial sets, identification does not involve removal of elements, but only modifications of operators). f: a minimal subdivision of a Klein bottle, embedded into E^3, with self-intersections (the edge is symbolized by the thick line). This is a surface subdivision which can be represented using 2-G-maps; if we intend to model the resulting subdivision of E^3, we have to explicity represent intersections and resulting volumes using 3-G-maps. g: quasi-manifolds are not manifolds. Identification of faces (A,E,B) and (D,E,C) produces a quasi-manifold (up and right), which is not a manifold, since the neighbourhood of vertex E is not a ball nor a half-ball; a similar quasi-manifold can also be constructed by identifying two central points of the circular faces of a full cylinder (down right).

2.3 Discussion

Semi-simplicial sets are well-known objects in algebraic topology, and generalized maps can be deduced from simplicial sets (cf. section 3). Geometric objects that can be associated with these structures are well defined. Thus, numerous notions, methods and results of algebraic topology can be applied and used in geometric modeling, and experiments have shown that it is important for implementations, constructions and controls.

Semi-simplicial sets and generalized maps are combinatorial structures, from which basic data structures can be derived in a straigthforward way (for instance, simplices are implemented as records, containing pointers which correspond to boundary operators). Constraints of consistency of these data structures (thus pre and post-conditions of construction operations) are obviously deduced from the definitions of semi-simplicial sets and generalized maps [LaLi1] [BD] [Lie1].

Modeled objects are structured ones. It is thus possible to structure embedding information, and to control consistency between topology and embedding (when it is needed: cf. Figure 3.f). For instance, semi-simplicial sets linearly embedded correspond to simplicial complexes. [LaLi1] has studied the modeling of semi-simplicial sets embedded using triangular Bezier patches : control points are associated with simplices, in such a way that the data structure is minimal, and consistency is based upon relations deduced from boundary operators. Modelers have been designed in which generalized maps are linearly embedded [BDFL], or embedded in a hierarchical way [Bor]: hierarchized embeddings are very useful for many applications (e.g. geology [MLCF]). Complementary structures are often added when non connected objects are handled (e.g. inclusion trees [HCRR]).

Topological properties are defined on semi-simplicial sets and can be extended on generalized maps. Basic properties as incidence and adjacency relations are explicitly defined and represented in data structures, through boundary operators (in practice, inverse operators are also explicitly represented in order to get a minimal cost when accessing cell neighbours[1]). For instance for 3D object reconstruction from stereo images, [Som] employs a combinatorial structure for describing the topology of segmented images, and its matching method is based upon a parallel traversal of the two structures corresponding to a pair of images (a similar structure describing the topology of the 3D object is constructed during this traversal). Many interesting properties can be also computed by data structure traversals. For instance, the topological surface associated with a 2-dimensional generalized map is completely defined by its number of boundaries, its orientability and its genus (using Euler characteristic) [Gri] [Tut], which can be computed on the 2-G-map. Some properties (and algorithms for computing them) are defined for higher dimensions (e.g.

---

[1] It is useless for involutions, since an involution is its own inverse.

orientability), although there is no equivalent classification. More generally, homology groups [Ago] [Gib] can be computed on semi-simplicial sets; it is also possible to decide whether or not a semi-simplicial set belongs to a given class (homogeneous, regular, pseudo or quasi manifold, etc) by simple traversals [ElLi1].

Such topological properties are important for controlling handled objects, their constructions, for proving algorithms, etc. Since topological information is represented by combinatorial structures, these properties can be computed without numerical computing. Moreover, these properties are important in order to control computing complexity. For instance, [FBDFR] studies an algorithm for locating points in a 3D mesh. After numerical optimizations (and conversion of some numerical computations into combinatorial computations), this algorithm spend more than 90% of computation time in traversals; since modelled subdivisions are orientable, it is possible to employ a specialized data structure, reducing significantly the cost in space and time (cf. section 3).

Construction operations have been defined in algebraic topology and in geometric modeling: creating simplices or cells, identification, cartesian product (extrusion being a particular case), chamfering, cell fusion, splitting, etc. These operations are implemented into modelers, often in a straightforward way and with minimal complexity. Note that several operations are less complex on generalized maps, since some informations are implicit: cf. Figure 4. An other example is the following. Given an n-G-map $G = (D,\alpha_0,...,\alpha_n)$ which describes the topology of a cellular quasi-manifold Q, the dual of G, $G^d = (D,\alpha_n,...,\alpha_0)$, describes the topology of the dual of Q (i.e. a subdivision of the same space in which i-cells correspond to (n-i)-cells in the initial subdivision, preserving adjacency relations).

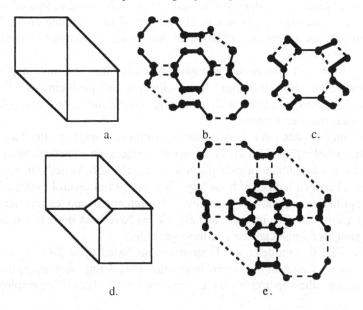

a.   b.   c.

d.   e.

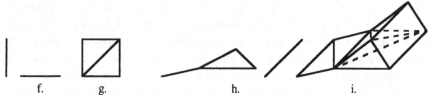

Fig. 4. a: a subdivision mainly composed by four faces; b. the corresponding 2-G-map (involution α_2 is not represented when it is equal to identity); c. 2 "copies" of the G-map describing the central vertex; d. chamfering the central vertex; e. the corresponding 2-G-map, obtained by inserting c. (chamfering operation can be easily defined for any cell dimension and any G-map dimension). f. two edges ; g. the 2D semi-simplicial set corresponding to their cartesian product; h: two semi-simplicial set, and i: their 3D cartesian product.

Previous structures, notions and operations are defined for any dimension, in an homogeneous manner. So, it is often easy to extend algorithms to higher dimensions. So, generic algorithms and operations can be defined. Sometimes, generalization to higher dimensions provides simplification of algorithms. This is not always true, since some properties are not applicable, for instance:

    – the Euler characteristic is defined for any semi-simplicial set (and thus for any generalized map), orientability can be computed on any generalized map, but the classification of 2-dimensional manifolds can not be generalized: cf. above.

    – A local ordering of edges around vertices exists for subdivisions of surfaces. This is generalized in 3D in a local ordering of faces around edges for 3D-manifolds, but the structure of edges around vertices is no more 1-dimensional, but 2-dimensional.

Extensibility for higher dimensions is important, since more and more applications initially concerned with (topologically) 2-dimensional problems (CAD/CAM, imagery[2], etc) are now concerned with (topologically) 3D and more recently 4D ones (in order to take time into account).

Semi-simplicial sets (and more generally simplicial sets) are the basis of an experimental modeler [LaLi1] [LaLi2]. Abstract simplices are embedded as triangular Bezier spaces, and assemblies of such patches can thus be constructed (since a natural relation exists between indices of boundary operators and of control points). Several classical topological operations have been implemented and experimented, as identification, cut, cartesian product, and algorithms have been designed to compute homology groups of simplicial sets and their generators.

Topofil [BD] is a modeler of 3D subdivisions based upon 3-G-maps linearly embedded into E^3. Classical operations in geometric modeling were adapted on 3-G-maps, but many other operations were conceived since Topofil is employed for

---

[2] For handling the topology of surface subdivisions, the first data structure used in topology-based modeling was conceived for computer vision problems.

different uses and application fields, as rendering, discrete geometry (for constructing discrete objects and checking their properties), animation, medical drawings, architecture, etc. The kernel of Topofil has been used for developing modelers in geophysics, meshes, etc. With a software engineering point of view, these extensions were simplified since Topofil has been algebraically specified, increasing thus reliability and reusing [BDFL].

3 Complexity

3.1 Reduced structures

For many applications, very efficient data structures are needed due to data size[3]. According to the properties of the modelled subdivisions, reduced simplicial and cellular structures can be derived from semi-simplicial sets. Examples are the following (for a more complete study, see [FuLi], [ElLi1], [ElLi2]):

   – An n-dimensional cubic set [Ser] is a collection of sets $K = (K_m)_{m=0,...,n}$ with operators (cf. Figure 5.a and 5.b):

$$\varepsilon^i_j: K_m \to K_{m-1}, \ m \geq 1, \ 1 \leq i \leq m, \ 0 \leq j \leq 1,$$

satisfying the following relations:

$$\varepsilon^i_j \varepsilon^k_l = \varepsilon^k_l \varepsilon^{i-1}_j \ \text{with} \ i > k.$$

K_m is a set of m-dimensional abstract cubes. 2m operators act on an m-cube, since an m-cube can be defined as the cartesian product of m edges, each one corresponding to a direction, and an edge has two extremities; so, ε^i_0 (resp. ε^i_1) gives the (m-1)-cube at the origin (resp. extremity) along the ith direction. Notions and operations (similar to that related with semi-simplicial sets) can be defined on cubic sets. Semi-simplicial sets can be naturally associated with cubic sets, through cartesian product. So, for "cubic subdivisions" (i.e. subdivisions in which cells are isomorphic to cubes), the cost of cubic sets is lower than that of the associated simplicial set (cf. figure 5.c; this is similar for cubic sets and associated generalized maps).

   – Simplicial quasi-manifolds make a sub-class of semi-simplicial sets (cf. section 2.2). Generalized maps correspond to cellular quasi-manifolds, which are defined as simplicial quasi-manifolds where simplices are structured into cells according to a numbering of vertices [Lie2] (this is closely related to the notion of barycentric triangulation): cf. figure 5.d and 5.e;

   – An n-dimensional combinatorial map (or n-map [Edm] [Jac] [Cor] [Lie1]) M is a set D with bijections (cf. Figure 5.f and 5.g):

$$\beta_i: D \to D, \ 1 \leq i \leq n,$$

such that:

$$\beta_1 \ \text{is a permutation on D}, \ \beta_i \ \text{is an involution on D}, \ 2 \leq i \leq n;$$
$$\beta_i \beta_j \ \text{is an involution}, \ 1 \leq i < i+1 < j \leq n.$$

---

[3] See also the study in [BF] [FB] about the complexity of 2D and 3D subdivisions.

Given a connected orientable n-G-map without boundaries $G = (D, \alpha_0, ..., \alpha_n)$, two n-maps M_1 and M_2 can be associated with G, each one corresponding to a possible orientation of G, i.e. : let d be a dart of D; $M_1 = (D_1, \beta_1, ..., \beta_n)$ where $\beta_i = \alpha_0 \alpha_i / D_1$, $1 \le i \le n$, and D_1 is the set of all darts which can be reached starting from d by successively applying β_i ($1 \le i \le n$) (the definition of M_2 is similar, and $d\alpha_0$ belongs to M_2). In practice, data structures deduced from n-maps are used for handling orientable cellular quasi-manifolds, with a lower space complexity than n-G-maps, since one dart of an n-map corresponds to two darts of the associated n-G-map.

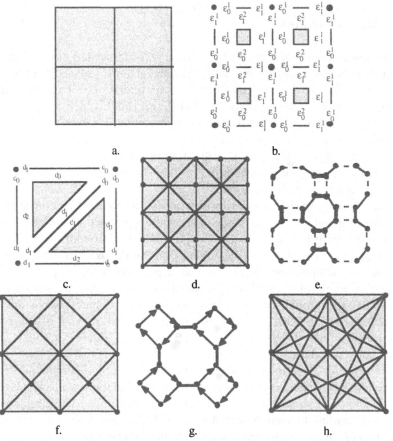

Fig. 5. a: a cubic subdivision. b: a corresponding cubic set. c: a semi-simplicial set associated with a 2-cube of b. d. the barycentric triangulation of a.; each cell is split by inserting a vertex at its barycenter, numbered with the cell dimension. e: the corresponding 2-G-map (involution α_2 is not represented when it is equal to identity); a dart is associated with each simplex, and involutions are deduced from boundary operators and vertex numbering. f: since the subdivision of d. is orientable, simplices can be merged two by two. g. the corresponding 2-map, which defines an orientation of d. h. a subdivision of a. where cells are not regular.

Other structures can be deduced from sub-classes of semi-simplicial sets (even cellular structures, defined as structured semi-simplicial sets), according to some mechanisms which employ the properties of sub-classes in order to reduce the information explicitly represented, e.g.:

    – when cells are cartesian products of simplices, the corresponding triangulation of cells is not explicitly represented in cubic or simploidal sets [FuLi];

    – chains [ElLi2] are deduced from semi-simplicial sets in which simplices are structured into cells (chains are associated with cellular complexes) and G-maps can be deduced from chains representing cellular manifolds (it is useless to explicitly represent 0-,...,(n-1)-cells, and the related boundary operators, when handling manifolds ; this is also true for simplicial manifolds).

Other mechanisms exist for other properties [ElLi1]. Main interest here consists in reducing the space complexity of data structures (information becomes implicit), and advantages presented before remain, since conversions between structures do not involve loss of information. For instance, for some very regular subdivisions as presented in figure 5, the explicit representation of all topological information is useless for many algorithms (even cubic sets are not optimal). Simplicial sets, cubic sets, etc., are needed in order to handle irregular assemblies of regular cells. Chains, generalized maps, etc. make it possible to handle irregular assemblies of irregular cells.

Operations for computing topological properties and construction operations have to be adapted to reduced structures. Time complexity of these operations can thus be strongly reduced (for instance, time complexity of some modification operations is reduced since less information has to be explicitly modified, for instance sewing or chamfering G-maps) or increased (when it is necessary to explicitly compute information being implicit in the handled structure). No general results exist, and it is necessary to carefully analyse space and time complexity according to the type of objects and operations which are needed for a particular use (partial results are presented in [ElLi1]).

In order to handle complex assemblies of free-form spaces, a modeler of simploids has been developed [FuLi], since natural links which exist between simplicial sets and control points of free-form spaces can be extended for simploidal sets and general simplicial algorithms.

[BeLi] studies the conception of Multifil, a modeler of cellular and simplicial complexes based upon the hierarchy of structures presented above, for handling general complexes with a minimal complexity in space and time. Hybrid structures are also studied, since few singularities exist in objects for many applications. For instance, usual objects are manifolds except in some few parts; it is thus useless to employ a general structure for representing all parts of the object: a manifold structure can be used, enriched when it is necessary to represent "non manifold" parts (similar mechanisms are used in [Mar]).

3.2 Hierarchized embeddings

For some applications, objects are structured into different levels, for representing informations related to the objects (for instance, a geographic map) or for optimizing time or space complexity (for instance different resolution levels can be distinguished in order to decrease time complexity of algorithms). Several approaches are based upon the natural recursivity of the definition of embedding [Bor] (a cell can be embedded onto a part of geometric space which is itself subdivided).

The key point here is consistency between topology and embedding. Since a discrete (or hierarchized) embedding is a "combinatorial embedding", its properties can be exactly defined and preserved through constructions. For instance, combinatorial maps are employed by [BrGu] [Bru] in order to handle segmented images: a 2-dimensional map is embedded into Z^2 by associating an "inter-pixel" path with each edge; these paths are the frontiers of regions, associated with faces, and their intersections are associated with vertices. For reconstructing 3D objects from stereo images, [Som] uses two 2-dimensional maps in order to describe the topology of a pair of segmented images, and an other 2-map describes the topology of the 3D object. This map is embedded into E^3, but also into the two other maps in order to control consistency between the object and its images (and to control its construction).

Problems related to consistency in a multi-level hierarchized embedding were studied by [Fey]. He proposed a definition of embedding of simplicial (and cellular) structures into lower-level simplicial (resp. cellular) structures (the lowest level being a regular subdivision of the 2D and 3D space, vertices being embedded onto points of Z^2 and Z^3), and of some properties of embedding which are necessary for applying some operations (e.g. boolean operations), in such a way that their results can be propagated to higher or lower levels.

4 Conclusion

Numerous combinatorial structures have been defined for representing regular or irregular subdivisions of geometric objects, for any dimension, and many construction operations and operations for computing topological properties have been proposed and experimented.

Using these structures, it is possible to handle "discretizations" of geometric objects, since they define the topology of structured continuous or discrete objects. Since they are combinatorial ones, data structures and algorithms have been defined without any loss of properties, and several modelers have been designed for different application fields as CAD, animation and image synthesis, architecture, digital imagery, etc.

Among numerous research directions, I think a main one is the following: Several works have shown that some results of topology-based geometric modeling can be useful for Discrete Geometry and Computer Imagery. A careful study of the usefulness of structures and algorithms has to be made, i.e. I think it is important to systematically study classical algorithms in Discrete Geometry and Imagery in order to get a definitive conclusion about the possible interests of Topology-based Modeling methods for these fields (thus following approaches of [Fio], [Lac], etc.). A step could be the conception of a modeler of discrete objects. It is thus important to carry on investigating algebraic topology in order to select and experiment constructions and properties which can be useful for geometric modelers and algorithms in Discrete Geometry.

Acknowledgements

I wish to thank (ex-)members of the Computer Science Research Center at Strasbourg, mainly J. Françon and J.P. Réveillès, Y. Bertrand, P. Borianne, H. Elter, C. Fey, L. Fuchs and V. Lang.

References

[Ago] Agoston, M.: Algebraic Topology: A First Course. Pure and Applied Mathematics, Marcel Dekker Ed., New-York, U.S.A. (1976)
[AFF] Ansaldi, S., de Floriani, L., Falcidieno, B.: Geometric Modeling of Solid Objects by Using a Face Adjacency Graph Representation. Computer Graphics **19**,3 (1985) 131–139
[Ale] Alexandroff: Combinatorial Topology. Graylock Press, Rochester, New-York, USA (1957)
[ArKo] Arquès, D., Koch, P.: Modélisation de solides par les pavages. Proc. of Pixim'89, Paris, France (1989) 47–61
[Bau] Baumgart, B.: A Polyhedron Representation for Computer Vision. AFIPS Nat. Conf. Proc. **44** (1975) 589–596
[BD] Bertrand, Y., Dufourd, J.-F.: Algebraic specification of a 3D-modeler based on hypermaps. CGVIP, **1** (1994) 29–60
[BDFL] Bertrand, Y., Dufourd, J.-F., Françon, J., Lienhardt, P.: Algebraic Specification and Development in Geometric Modeling. Proc. of TAPSOFT'93 Orsay, France (1993)
[BeLi] Bertrand, Y., Lienhardt, P.: Spécification algébrique et sortes ordonnées pour un multi-modeleur d'extensions des cartes. Journées du GDR/PRC de Programmation, Orléans, France (1996).
[BF] Bertrand, Y., Françon, J.: 2D manifolds for Boundary Representation: A statistical study of the cells. Research Report R97-8, Université Louis Pasteur, Strasbourg, France (1997)
[Bor] Borianne, P.: Conception of a Modeller of Subdivisions of Surfaces based on Generalized Maps. PhD Thesis **037**, Université Louis Pasteur, Strasbourg, France (1991)
[Bra] Braid, I .: The Synthesis of Solids Bounded by Many Faces. Communications of the A.C.M. **18**,4 (1975) 209-216

[BrGu] Braquelaire, J.-P., Guitton, P.: A Model for Image Structuration. Proc. of Computer Graphics International'88 Genève, Switzerland (1988)

[Bri] Brisson, E.: Representing Geometric Structures in D Dimensions: Topology and Order. Discrete and Computational Geometry 9 (1993) 387–426

[Bru] Brun, L.: Segmentation d'images à base topologique. PhD Thesis 1651, Université de Bordeaux I, Bordeaux, France (1996)

[BrSi] Bryant, R., Singerman, D.: Foundations of the Theory of Maps on Surfaces with Boundaries. Quart. Journal of Math. Oxford 2,36 (1985) 17–41

[CCM] Cavalcanti, P.R., Carvalho, P.C.P., Martha, L.F.: Non-manifold modeling: an approach based on spatial subdivision. Computer-Aided Design 29, 3 (1997) 299–220

[Cor] Cori, R. Un code pour les graphes planaires et ses applications. Astérisque 27 (1975)

[CrRe] Crocker, G., Reinke, W.: An Editable Non-Manifold Boundary Representation. Computer Graphics and Applications 11,2 (1991)

[DoLa] Dobkin, D., Laszlo, M.: Primitives for the Manipulation of Three-Dimensional Subdivisions. Proc. of 3rd Symposium on Computational Geometry Waterloo, Canada (1987) 86–99

[Edm] Edmonds, J.: A Combinatorial Representation for Polyhedral Surfaces. Notices Amer. Math. Soc. 7 (1960)

[ElLi1] Elter, H., Lienhardt, P.: Different Combinatorial Models based on the Map Concept for the Representation of Different Types of Cellular Complexes. IFIP TC5/WG5.10 Working Conference on Geometric Modeling in Computer Graphics, Genova, Italy (1993)

[ElLi2] Elter, H., Lienhardt, P.: Cellular complexes as structured semi-simplicial sets. Int. Journal of Shape Modeling, 1,2 (1994) 191–217

[FB] Françon, J., Bertrand, Y.: 3D manifolds for Boundary Representation: A statistical study of the cells. Research Report R96-9, Université Louis Pasteur, Strasbourg, France (1997)

[FePa] Ferruci, V., Paoluzzi, A.: Extrusion and Boundary Evaluation for Multidimensional Polyhedra. Computer-Aided Design 23,1 (1991) 40–50

[Fey] Fey, C.: Etude de plongements hiérarchisés de subdivisions simpliciales ou cellulaires. PhD Thesis, Université Louis Pasteur, Strasbourg, France (1996)

[Fio] Fiorio, C.: A Topologically Consistent Representation for Image Analysis: The Frontiers Topological Graph. Proc. of DGCI'96, Lyon, France (1996) 151–162

[FBDFR] Fousse, A., Bertrand, Y., Dufourd, J.-F., Françon, J., Rodriguès, D.: Localisation des points d'un maillage généré en vue de calculs en différences finies. Journées "Modélisation du sous-sol". Orléans, France (1997).

[Fra] Françon, J.: On Recent Trends in Discrete Geometry in Computer Science. Proc. DGCI'96, Lyon, France (1996) 3–16

[FrPi] Fritsch, R., Piccinini, R. A.: Cellular Structures in Topology. Cambridge University Press (1990)

[FuLi] Fuchs, L., Lienhardt, P.: Topological Structures for d-Dimensional Free-Form Objects. CAGD'97, Lillehammer, Norway (1997)

[GHPT] Gangnet, M., Hervé, J.-C., Pudet, T., Van Thong, J.-M.: Incremental Computation of Planar Maps. Computer Graphics 23,3 (1989) 345–354

[Gib] Giblin, P.J.: Graphs, surfaces and homology. Chapman and Hall, London, UK (1977)

[Gri] Griffiths, H.-B. Surfaces. Cambridge University Press Cambridge, U.K. (1981)

[GuSt] Guibas, L., Stolfi, J.: Primitives for the Manipulation of General Subdivisions and the Computation of Voronoï Diagrams. A.C.M. Transactions on Graphics **4,2** (1985) 74–123

[GCP] Gursoz, E. L., Choi, Y., Prinz, F. B.: Vertex-based Representation of Non-Manifolds Boundaries. in Geometric Modeling for Product Engineering M. Wozny, J. Turner and K. Preiss eds., North-Holland (1989) 107–130

[HC] Hansen, O.H., Christensen, N.J.: A model for n-dimensional boundary topology. Proc. 2nd ACM Symp. Solid Modeling Foundations and CAD/CAM Applications (J. Rossignac and J. Turner eds) Montréal, Canada (1993) 65–73

[HCRR] Halbwachs, Y., Courrioux, G., Renaud, X., Repusseau, P.: Topological and Geometric Characterization of Fault Networks Using 3-Dimensional Generalized Maps. Mathematical Geology, **28, 5** (1996) 625–656

[Jac] Jacque, A.: Constellations et Graphes Topologiques. Colloque Math. Soc. Janos Bolyai (1970) 657–672

[KIE] Kenmochi, Y., Imiya, A., Ezquerra, N.: Polyhedra generation from lattice points. Proc. DGCI'96, Lyon, France (1996) 127–138

[Lac] Lachaud, J.-O.: Topologically Defined Isosurfaces. Proc. of DGCI'96, Lyon, France (1996) 245–256

[LaLi1] Lang, V., Lienhardt, P.: Geometric Modeling with Simplicial Sets. Proc. of CGI'96, Pohang, Korea (1996)

[LaLi2] Lang, V., Lienhardt, P.: Cartesian Product of Simplicial Sets. Proc. of WSCG'97, Plzen, Czech Republic (1997)

[Lie1] Lienhardt, P.: Topological Models for Boundary Representation: a Comparison with N-Dimensional Generalized Maps. Computer-Aided Design **23,1** (1991) 59–82

[Lie2] Lienhardt, P.: N-Dimensional Generalized Combinatorial Maps and Cellular Quasi-Manifolds. Int. Journal of Computational Geometry and Applications **4,3** (1994) 275–324

[LuLu] Luo, Y., Lukacs, G. A.: A Boundary Representation for Form-Features and Non-Manifold Solid Objects. Proc. of 1st ACM/Siggraph Symposium on Solid Modeling Foundations and CAD/CAM Applications Austin, Texas, U.S.A. (1990)

[Män] Mäntylä, M.: An Introduction to Solid Modeling. Computer Science Press Rockville, U.S.A. (1988)

[May] May, J. P.: Simplicial Objects in Algebraic Topology. Van Nostrand Princeton (1967)

[Mar] Marcheix, D.: Modélisation géométrique d'objets non-variétés: construction, représentation and manipulation. PhD Thesis, Université de Bordeaux I, Bordeaux, France (1996)

[Mil] Milnor: The geometric realization of semi-simplicial complexes. Ann. of Math. **65** (1957) 357–362

[MLCF] Mallet, J.-L., Levy, B., Conreaux, S., Fousse, A.: G-maps, a new topological model for gOcad. 15th Gocad Meeting, Nancy, France (1997)

[MuHi] Murabata, S., Higashi, M.: Non-manifold geometric modeling for set operations and surface operations. IFIP/RPI Geometric Modeling Conference Rensselaerville, N.Y. (1990)

[PBCF] Paoluzzi, A., Bernardini, F., Cattani, C., Ferrucci, V.: Dimension Independent Modeling with Simplicial Complexes. ACM Transactions on Graphics, **12, 1** (1993)

[RoOC] Rossignac, J., O'Connor, M.: SGC: A Dimension-Independent Model for Pointsets with Internal Structures and Incomplete Boundaries. in Geometric Modeling for Product Engineering M. Wozny, J. Turner and K. Preiss eds., North-Holland (1989) 145–180

[Ser] Serre,J.-P.: Homologie singulière des espaces fibrés. Ann. of Math. **54** (1951) 425–505

[SeTh] Seifert, H., Threlfall, W.: A textbook of topology. Academic Press New York, (1980)

[Som] Sommellier, L.: Mise en correspondance d'images stéréoscopiques utilisant un modèle topologique. PhD Thesis, Université Claude Bernard, Lyon, France (1997)

[Spe] Spehner, J.-C.: Merging in Maps and Pavings. Theoretical Computer Science **86**, (1991) 205–232

[Tak] Takala, T.: A taxonomy of geometric and topological models. Computer Graphics and Mathematics (B. Falcidieno, I. Herman and C. Pienovi eds.), Springer (1992)

[Tut] Tutte, W. Graph Theory. Encyclopaedia of Mathematics and its Applications, Addison Wesley, Menlo Park, U.S.A. (1984)

[Vin] Vince, A.: Combinatorial Maps. Journal of Combinatorial Theory Series B **34** (1983) 1–21

[Wei1] Weiler, K.: Edge-based Data Structures for Solid Modeling in Curved-Surface Environments. Computer Graphics and Applications **5,1** (1985) 21–40

[Wei2] Weiler, K.: The Radial-Edge Data Structure: A Topological Representation for Non-Manifold Geometry Boundary Modeling. Proc. IFIP WG 5.2 Working Conference Rensselaerville, U.S.A. (1986), in Geometric Modeling for CAD Applications Elsevier (1988) 3–36

2D Recognition

Applications of Digital Straight Segments to Economical Image Encoding

Vladimir Kovalevsky

Technische Fachhochschule Berlin
Luxemburger Str. 10, 13353 Berlin, Germany
email: kovalev@tfh-berlin.de

Abstract. A new classification of digital curves into boundary curves and visual curves of different thickness is suggested. A fast algorithm for recognizing digital straight line segments in boundary curves is presented. The algorithm is applied to encode the boundaries of homogeneous regions in digitized images. The code is economical and enables an exact reconstruction of the original image.

1 Introduction

The aim of this presentation is to demonstrate that methods of recognizing, encoding and decoding segments of digital straight lines may be successfully applied to economical encoding of digitized images and to exactly reconstruct the original image from the code. Methods of recognizing digital straight segments (DSS) are known during a long time. One of the first methods is due to Freeman [Fre74]. He suggested to analyze the regularity in the pattern of the directions in the chain code [Fre61] of a digital curve. Anderson and Kim [AndKim85] have presented a deep analysis of the properties of the DSS's and suggested a different algorithm based on calculating the convex hull of the points of the digital curve to be analyzed. The author has suggested a simple and fast version of this algorithm [Kov90] and demonstrated that the recognition may be performed as a successive, point for point, actualization of the coefficients of the linear inequality, which all points of a DSS must satisfy. Today this inequality is generally known. It was successfully used in many recent works, e.g. [RevDeb94], [Deb95], [Franc96].

The author has devoted his investigation to a special class of the DSS's: the boundary lines. The reader will find in this presentation a *new classification of digital curves* into boundary curves and visual curves. Boundary curves and lines are a useful means for fast drawing of regions defined by their boundaries. The algorithm for filling the interior of a closed boundary curve represented as a sequence of boundary cracks [Kov84], [Kov94a] is very simple and fast. It is an important tool for the reconstruction of images encoded by means of the boundaries of homogeneous regions. However the known methods of encoding boundaries without loss of information need relatively much memory space. We suggest here a *new method of encoding* digital straight segments without loss of information which is more economical then the known methods.

Sections 2 and 3 contain a short summary of the topological and geometrical background of the further sections. Section 4 is devoted to the new classification of digital curves. It also contains a short description of a new fast algorithm for drawing visual curves of arbitrary width with antialiasing when given an equality of the curve in the form $F(x,y)=0$. $F(x,y)$ may be given as any floating point function in a programming language.

Section 5 presents a new derivation of the well-known properties of the DSS's. This section is the theoretical foundation of the recognition algorithm. The algorithm described in Section 6 is not new: this is the algorithm of [Kov90]. What is new is the method of encoding the parameters of the DSS's which is presented in Section 7. Section 8 contains some experimental results.

2   Topological Background

We consider here the digital plane as a two-dimensional cell complex [Kov89] rather then a set of pixels. Thus our digital plane contains *cracks* and *points* besides the pixels. Cracks are the sides of the pixels, the latter being considered as square areas. From the point of view of cell complexes pixels are two-dimensional and cracks are one-dimensional cells. The points are end points of the cracks and therefore the corners of the pixels. Points are zero-dimensional cells.

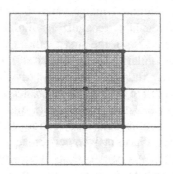

Fig. 1. Example of a two-dimensional complex
with the crack boundary of the shaded subset

As demonstrated in [Kov89], considering the plane as a cell complex brings many advantages: there are no more connectivity paradoxes, a boundary becomes a thin curve with a zero area, the boundary of a region and that of its complement are the same etc. The definition and the processing of digital curves and especially that of digital straight lines becomes simpler and clearer. The most important advantage from the point of view of economical encoding and exact reconstruction of images is the possibility to fill the interior of crack boundaries by an extremely simple and fast algorithm [Kov84], [Kov94a] which cannot be applied when representing boundaries as sets of pixels.

The rest of this section contains a short summary of the topological notions important for this presentation. Please refer to [Kov89] for more details and topological foundations. The reader acquainted with cell complexes may skip the rest of this section.

An D-dimensional cell complex is a structure consisting of abstract elements called cells. Each cell is assigned an integer value from 0 to D called its dimension. There is a bounding relation imposed onto the cells: a cell of a lower dimension may bound some cells of a higher dimension. An example of a two-dimensional complex is shown in Fig. 1.

The pixels are represented in Fig. 1 as the interiors of the squares, the cracks as the sides of the squares and the points, i.e. the 0-cells, are the end points of the cracks and simultaneously the corners of the pixels.

Now let us introduce some notions which we shall need in the sequel. A *boundary crack* of a subset S of a complex is a crack separating a pixel belonging to S from another pixel not belonging to S. The boundary cracks of the shaded subset in Fig. 1 are drawn as bold lines. The *boundary* (also known as *crack boundary*, [RosKaak82]) of a subset S is the set of all boundary cracks of S and all end points of these cracks. A boundary contains no pixels and is therefore a „thin" set whose area is zero. A connected subset of a boundary is called a *boundary curve*. For the notion of connectedness please refer to [Kov89].

3 The Coordinates

We consider the digital plane as a *Cartesian two-dimensional complex* [Kov94b], i.e. as a Cartesian product of two one-dimensional complexes which are the coordinate axes of the plane. The X-coordinate is the row number, the Y-coordinate is the line number. We use here the coordinate system of computer graphics, i.e. the positive X-axis runs from left to right, the positive Y-axis runs from top to bottom.

We often need Euclidean coordinates to discuss problems of digitizing Euclidean objects. To remain in the frame of cell complexes we suggest to consider Euclidean coordinates as rational numbers with a relatively great denominator. This corresponds to any computer model since the floating point variables in computers are rational numbers with great denominators.

It is possible to introduce additionally to the complex of the digital plane another cell complex, the *fine complex*, whose cells are obtained by subdividing the original cells into many smaller cells. The coordinates in the fine complex are rational numbers. Both coordinate systems of the original and the fine complex may be adjusted to each other in such a way that the 0-cells (i.e. the points) of the original complex have integer coordinates. Then we have the possibility to consider other points lying inside the pixels and cracks. These point have fractional coordinates. Particularly important for this presentation are the points in the middle of the pixels. They have „half-integer" coordinates, i.e. fractional coordinates with an odd numerator and the denominator equal to 2: e.g. 0.5, 1.5, 2.5 etc.

4 Classification of Digital Curves

Curves are considered in the classical geometry as objects with zero width and zero area. In digital geometry we consider the number of pixels (being elementary areas) in a subset S as the measure of the area of S. A boundary curve is a sequence of cracks. It contains no pixels and thus it exactly corresponds to the classical notion of a curve. Although a boundary curve contains no pixels, it is possible to make it visible on the screen of a computer. If we are ready to make a magnified representation of a complex, then we draw a 2-cell as a small square consisting of $N \times N$ screen pixels. The cracks can be represented as thin vertical or horizontal bars one screen pixel wide and N screen pixels long. A digital curve is then a sequence of such vertical an horizontal bars. If one does not want to magnify the complex then each crack of a curve may be represented by a single screen pixel whose coordinates correspond to the upper or to the left end of the crack. Such a representation has two drawbacks:

1. The image of the curve is slightly displaced relative to the positions of pixels which may be simultaneously represented on the screen;
2. The image of the curve is too thick at some places where an upper end point of a vertical crack is simultaneously the left end point of a horizontal crack. The image of the curve is not a „thin" sequence of screen pixels. It is not a skeleton.

To overcome these drawbacks we suggest here to consider in computer graphics two classes of curves: the *boundary curves* as sequences of cracks and the *visual curves* as sequences of pixels. Boundary curves are important as a means to exactly specify the boundaries of subsets. They are a most suitable tool to draw areas filled by some color: the filling algorithm of [Kov94a] being applicable in this case is much simpler and faster then any algorithm suitable to fill the interior of a closed sequence of pixels. The exact visual representation of boundary curves is less important.

Visual curves must be used in the cases when a Euclidean curve, e.g. specified by its equation, must be displayed in such a way that it gives a visual impression of the mathematically specified curve as exact as possible, including the impression of homogeneous thickness. For the last purpose the so called antialiasing methods have been developed in computer graphics. The author has developed a new universal algorithm for drawing visual curves of arbitrary width with antialiasing. The algorithm may draw any curve specified by an equation of the form $F(x,y)=0$. The equation may be given as a function in a programming language while having floating point arguments and returning a floating point value. The algorithm implicitly calculates two curves displaced from $F(x,y)=0$ by the half of the desired width along and against the gradient of $F(x,y)$ and estimates the area of the part of each pixel (in the vicinity of $F(x,y)=0$) which part lies between the two displaced curves. The color of the pixel to be drawn depends upon the estimated area. The estimation is realized without use of a finer raster (as e.g. in [Whit83]) and is therefore very fast and precise.

In the following sections we consider only boundary curves and a method of subdividing a digital curve in digital straight line segments of maximal possible length. The segments are also boundary segments.

5 Properties of Digital Straight Segments

We are speaking here about digital straight segments (DSS) rather then about digital lines because in practice we always consider finite subsets of the digital plane. Such a subset can only contain a subset of a line, i.e. a line segment.

It is natural to define a digital straight segment as a result of the digitization of a Euclidean straight segment. Various digitization schemes of digitizing curves were suggested (e.g. [AndKim85]). Most of them are based on testing the crossings of the given curve with the grid lines. It is rather difficult to mathematically justify these schemes or specify some reasons for preferring one of them. Therefore we are suggesting here one scheme more which follows from the commonly used practical way of digitizing regions in raster images.

The natural way of digitizing a region in the Euclidean plane consists in subdividing the plane in regularly spaced elementary areas corresponding to the pixels and in measuring for each pixel its portion covered by the given Euclidean region. The measured value must be compared with a threshold. A pixel must be set to 1 if the measured portion is greater than the threshold and set to 0 otherwise. We call this method *digitizing by thresholding*.

This scheme may be applied to *digitizing Jordan curves* in the following way:

1) Given a Euclidean Jordan curve JC construct the Euclidean region R which is the interior of JC;
2) Digitize R by thresholding and place the results into the 2-dimensional cell complex thus defining a digital region DR;
3) Find the boundary of DR. This is the digital image of JC.

The above scheme is slightly bothersome because of the necessity to measure area portions for each pixel. There is however a simple particular case of the scheme. Consider first the digitization of a straight line. The corresponding region R is a half-plane H. Let us digitize H with a threshold equal to 0.5. Then digitization may be performed *without calculating area portions* since a half-plane covers more than 0.5 of a rectangular elementary area iff the middle point of the area is inside H which may be found by substituting the coordinates of the middle point into the inequality of H. The same scheme may be used for digitizing curves with restricted curvature. A simple calculation shows that the error in estimating the area portion is less than 1% if the curvature radius is greater than 6 times the side of a pixel.

Thus we arrive at the following digitization scheme for Euclidean curves (e.g. given by equations):

1) Subdivide the plane in regularly spaced squares and define the middle point of each rectangle;
2) Establish a mapping of the set of the squares onto the 2-dimensional complex while mapping the squares onto the 2-cells (pixels) and their sides onto the 1-cells (cracks).
3) Find two adjacent pixels whose middle points lie on different sides of the given Euclidean curve. If the curve is given by an equation like $F(x,y)=0$, then the values of $F(x,y)$ at the said two middle points must have different signs. The crack between the two pixels is then a boundary crack. It belongs to the digital curve.

The well-known properties of DSS's (see e.g. [AndKim85]) may be easily derived from our digitization scheme as follows. Consider a half-plane whose inequality is $H(x,y)=P \cdot x+Q \cdot y+R>0$, where P and Q are integers. The middle points of all pixels are subdivided into two subsets: those with positive and those with non-positive values of $H(x,y)$. Let us call them „positive" and „negative" middle points. For every 0-cell C belonging to the boundary of these subsets (Fig. 2) there exist four adjacent middle points whose distance to C is equal to $\sqrt{2}/2$ of the length of a pixel side. We shall consider since now the length of a pixel side as the unit of measurement. The four middle points compose two diagonal pairs (MP_1, MP_4) and (MP_2, MP_3) each having the 0-cell C in the middle. These pairs are shown in Fig.2. Consider that of the two pairs for which the absolute difference of the values of $H(x,y)$ is greater (or equal) to that for the other pair (the absolute differences are equal only for half-planes with horizontal or vertical boundaries). Let us call the corresponding diagonal direction the *main diagonal* for H. It is (MP_2, MP_3) in Fig. 2.

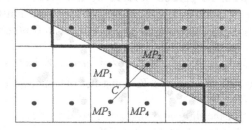

Fig. 2. The positive and the negative middle points of a half-plane
and the main diagonal (MP_2, MP_3) of the half-plane

Since the vector connecting the middle points of a pair has both its X- and Y-components equal to ± 1, it is easy to see that the maximum absolute difference is equal to $|P|+|Q|$. The values of the points belonging to the same pair must have different signs or, more exactly, one value must be positive and the other negative or zero. Thus the maximum positive value of $H(x,y)$ for all „positive" middle points adjacent to a boundary point like C cannot be greater than $|P|+|Q|$ and the adjacent „negative" middle points have all values strongly greater than $-(|P|+|Q|)$. Since the 0-cells of the boundary lie exactly in the middle between the middle points of each mean diagonal pair, the values of $H(x,y)$ for the 0-cells must satisfy the condition:

$$-(|P|+|Q|)/2 < H(x,y) \leq +(|P|+|Q|)/2. \tag{1}$$

This is an important property of the points of a DSS which may be used for fast recognition of the DSS's.

The next important property of a DSS is its periodicity. Consider a 0-cell C with coordinates (X,Y) belonging to the DSS which belongs to the boundary of the half-plane $H(x,y)=P \cdot x+Q \cdot y+R \geq 0$. Another 0-cell with coordinates $(X+a,Y+b)$ will have the same value of $H(x,y)$ if the condition $P \cdot a+Q \cdot b=0$ holds. Let $P=P_s \cdot GCD$ and $Q=Q_s \cdot GCD$ where GCD is the greatest common divisor of P and Q. Then the smallest values of a and b must satisfy the condition $P_s \cdot a + Q_s \cdot b = 0$ or $a = Q_s$ and $b = -P_s$. Therefore the vector (a, b) defines the period of the DSS. It is not correct to affirm that every DSS is periodic: it may consist of exactly one period. In such a case the assertion about the periodicity has no sense. A correct assertion is: any DSS has a subsequence, called its period, such that any multiple repetition of the period always produces a DSS. These properties enable the recognition of DSS's: given a digital curve CV as a sequence of 0-cells and cracks it is possible to find the longest segment of CV which is a DSS.

Let us remind some definitions introduced in [Kov90], namely that of the base of a DSS and of its starting and end points. Consider the „negative" and „positive" subsets of pixels corresponding to a given half-plane inequality $H(x,y) \geq 0$ as mentioned above. A connected subset of the boundary of these subsets is a DSS. It consists of cracks and points and contains no pixels. The Euclidean straight line containing the longest edge of the convex hull of the points is called the *base of the DSS* (the same as the „nearest support" of [AndKim85]). There is another base touching the hull on the opposite side and being parallel to the first one.

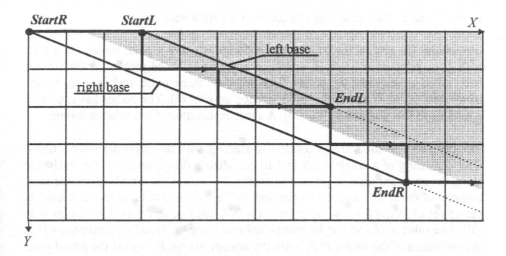

Fig.3. The bases of a DSS and their end points

Let us give the DSS such an orientation (arrows in Fig. 3) that the „positive" pixels are to the left hand side of the DSS. Correspondingly we call one of the bases the left and the other the right base. In Fig. 3 the positive half-plane is shown as the shaded region. The cracks of the corresponding DSS are drawn as bold lines. *StartL* and *EndL* are the vectors of the end points of the left base; *StartR* and *EndR* are those of the right base.

Let us choose the constant R in the inequality $H(x,y) = P \cdot x + Q \cdot y + R \geq 0$ such that $H(x,y)$ be equal to zero for points lying on the right base. The difference between the minimum and the maximum values of $H(x,y)$ at all points of the DSS remains unchanged. Therefore according to (1) the values of $H(x,y)$ must be now in the range:

$$0 \le H(x,y) < |P|+|Q|. \tag{2}$$

If P and Q are relatively prime then the difference between the maximum value of $H(x,y)$ and $|P|+|Q|$ has the smallest possible integer value 1. Therefore for all points of a DSS the following inequality holds:

$$0 \le H(x,y) \le |P|+|Q|-1. \tag{3}$$

The values of P and Q may be considered as the components of the normal to the Euclidean line specified by $H(x,y)=0$. It is more convenient for our purpose to consider a vector along the base of the DSS whose components are relatively prime. The vector is obviously perpendicular to the normal and has the X-component $N=-Q_s$ and the Y-component $M=P_s$. Thus we can rewrite the above inequalities as:

$$H(x,y)=M \cdot (x-X_r)-N \cdot (y-Y_r); \tag{4}$$

$$0 \le H(x,y) \le |M|+|N|-1; \tag{5}$$

where X_r and Y_r are coordinates of a point on the right base.

Inequalities (5) give us the possibility the decide whether a point belongs to a DSS with given parameters M, N, X_r and Y_r. A more complicated problem is the inverse one: given a sequence of points find whether there exist such values of the parameters M, N, X_r and Y_r that the sequence satisfies (5). The author has suggested a solution of this problem in [Kov90]. A short explanation of the solution follows.

The main idea of the solution consists in starting with the parameters for a trivial DSS consisting of a single crack and in actualizing the parameters after each step along the given sequence of points and cracks. To explain the actualization one more property of the DSS's must be demonstrated. We have showed that the difference of the values of $H(x,y)$ for any two points of a DSS must be less than $|M|+|N|$ (compare (5)). The value of $H(x,y)$ may be interpreted according to (4) as being proportional to the projection of the vector $P-P_r$ onto the normal, where $P=(x,y)$ is the actual point of the sequence to be tested and $P_r=(X_r, Y_r)$ is a point on the right base. If $H(x,y)$ at P becomes exactly equal to $|M|+|N|$ (or respectively to -1) while it was in the range of (5) for all other already tested points, then there is the possibility to slightly rotate the normal in such a way, that $H(x,y)$ at P becomes slightly less (respectively greater) thus satisfying the conditions (5) and the values of $H(x,y)$ for all other points remain in the permitted range. The author has proved that this rotation must be performed by moving the end point of one of the bases to the actual point P. This is the end point closest to P and belonging to the base closest to P. This idea is realized in the following algorithm.

6 Recognition of Digital Straight Segments

The recognition algorithm uses the inequality (5) of a DSS. We consider all points as vectors. We denote the components of a vector V as $V.X$ and $V.Y$. The pair of values (N, M) is also a vector denoted as $TANG$ („tangent"): $TANG.X=N$, $TANG.Y=M$. The instruction „continue" in the following description of the algorithm means „change nothing, get the next point of the sequence". The instruction „break" means „the actual point P does not belong to the DSS; store the point preceding P as the end point of the DSS; return a break message". We shall denote by $HL(x,y)=0$ the equation of the left and by $HR(x,y)=0$ that of the right base. The notations $StartL$, $EndL$ etc. were explained in the previous section (Fig. 3).

Take the first two points P_1, P_2 of the given curve and set $StartL=StartR=P_1$; $EndL=EndR=P_2$. Set the starting values of $TANG=EndL-StartL$ (M and N take the values of either to 0 or ±1). Set the half-plane values $HL=HR=0$.

1. Store the first two different directions of cracks appearing since the start.
2. For every next point P of the curve:
   2.1 Test the direction of the last step, whether it is equal to one of the two directions having appeared until now:
   If it is not, then break (the sequence is no more a DSS);
   else increase HL by $-(M \cdot STEP.X - N \cdot STEP.Y)$ and HR by $(M \cdot STEP.X - N \cdot STEP.Y)$, where $STEP$ is the unity vector having the direction of the last step. (Note that one of the components of $STEP$ is zero and the other is equal to ±1).
   2.2 Test the value of HR:

HR is positive:	continue;
HR is zero:	set $EndR=P$, continue;
HR is equal to -1:	calculate the new values of $TANG$, $EndR$, $StartL$, HL (the only two multiplications); set $HR=0$; continue;
HR is less than -1:	break.

   2.3 Test in a similar way the value of HL.
End of the algorithm.

After a "break" the point PP directly preceding the actual point P must be stored as the end point of the DSS. The next DSS begins at PP while PP is the first and P the second point of the new DSS and so on.

7 Encoding of curves

The Crack Code
A well-known way of encoding digital curves is that of the Freeman code [Fre61]. Boundary curves in a two-dimensional complex are sequences of cracks and points. Oriented cracks may have only four directions. Therefore they may be encoded by a Freeman code with four directions which is also well-known as the crack code (see e.g. [RosKaak82]).

This way of encoding curves is a rather economical one especially if only two bits per crack are used. Its main drawback is the difficulty of performing geometrical transformations: only a translation is easy to realize. Rotation and scaling are hardly realizable without converting the crack code into some other code more suitable for geometrical transformations.

End Points of DSS's

Another way of encoding curves consists in decomposing the curve into as long as possible DSS's and recording the coordinates of their end points. This code makes geometrical transformations easily realizable: it suffices to multiply the vectors corresponding to all end points with the matrix of the desired transformation, eventually in homogeneous coordinates. However this code does not allow an exact reconstruction of the original digital curve since there exist many different DSS's having the same end points. The distance between any two such DSS's is never greater than the pixel's diagonal. Therefore the difference between such two DSS's may be considered in many applications as negligible. In such cases this way of encoding curves is the simplest and the most economical one, especially if we record coordinate increments rather then the coordinates itself. If however an exact reconstruction is necessary than the following ways of encoding are possible.

Floating Point Coordinates

A DSS is uniquely specified by its end points and any Euclidean straight line which is a preimage of the DSS. The Euclidean line parallel to the bases and lying exactly in the middle between them is such a line. We shall call the line *the axis of the DSS*. The parameters of the axis and the location of the end points may be combined if we calculate the crossing points of the axis with the main diagonals (s. Section 5) containing the end points. The crossing points uniquely define both the axis and the integer end points which are the integer points nearest to the crossing points. The drawbacks of this way of encoding are as follows:

a) the memory demand is relatively great, e.g. two „float" coordinates of each of two crossing points with two axes of subsequent DSS makes $4\times2\times2$ bytes per end point;

b) a common end point of two subsequent DSS's of a digital polygon may be „split" after geometrical transformations into two different integer points thus making the polygon disconnected.

There are some rather complicated ways of overcoming one of the drawbacks but not both of them.

Additional Integer Parameters of a DSS

A DSS may be uniquely specified by its end points and by one of its bases. Specifying the base by the coordinates of its end points (which are mostly different from the end points of the DSS) is rather redundant. A more economical way of encoding consists in specifying two integers M and N for the direction of the base accompanied by the distance of the starting point of the DSS from the base. The distance itself is a small value and must be represented by a floating point variable. However there is a possibility to specify a positive integer L not greater than

$|M|+|N|-1$ which together with the values of M and N specifies the distance uniquely. The coordinates of the end points of the DSS may also be encoded economically: only the coordinates of the starting point of the curve must be specified explicitly; all other coordinates may be defined by means of the number NC of cracks in the current DSS. Thus we need four integers per DSS: NC, L, M and N. Since the majority of the DSS's are relatively short, these integers are mostly small ones: all four integers may be packed into a word of 2 bytes. For longer DSS's a longer word of 3 or 4 bytes may be needed, but the frequency of such long words is low. Therefore the average number of bytes per one DSS lies between 2 and 3.

8 Experiments

The author has developed a program which traces the boundaries of regions with constant gray levels in an image of 1 byte per pixel, dissects the boundaries in as long as possible DSS's and encodes them by the code described in the previous section. Another program reconstructs the image from the code. The reconstructed image is always identical with the original one. Numerous experiments have shown that the average number of bytes per one DSS is of order of 2.3. This leads to an economical encoding of images: the compression rate is of the order of about 20 to 40 for images with a low density of fine details and of about 3 for images with many fine details or many gray levels. The length of the code is approximately inversely proportional to the number of gray levels in the original image. To obtain a high compression rate the image must be properly smoothed and quantized.

Fig. 4. An example of an encoded image

Fig. 4 shows an image of 390×480=187200 pixels with 32 gray levels. It was encoded with 60080 bytes. Thus the compression rate was about 3.1.

When encoding the same images by means of the crack code, the compression rate is about 1.5 for a low density of details and about 0.15 (no compression) for images with many fine details. Thus the encoding by means of the DSS's brings a much higher compression rate than that of crack codes. The suggested method is less efficient then the well-known ARJ method, but our method represents the image in a „geometrical" way: i.e. it is possible to extract from the code the boundaries of the regions and thus perform image analysis by means of the code.

References

[AndKim85] T. A. Andersen and C.E. Kim: Representation of digital line segments and their preimages. CVGIP 30, 279-288 (1985).

[Deb95] I. Debled-Renesson: Etude et reconnaissance de droites et plans discrets. Thése de doctorat soutenue à l'Université Louis Pasteur de Strasbourg, 1995.

[Franc96] J. Françon, J.-M. Schramm and M. Tajine: Recognizing arithmetic straight lines and planes. In: Miguet, S., Montanvert A., Ubéda S. (eds.): Discrete Geometry for Computer Imagery. Berlin Heidelberg New York Tokyo: Springer 1996 (Lecture Notes in Computer Science, vol. 1176, pp. 141-150).

[Fre61] H. Freeman: On the encoding of arbitrary geometric configurations. IRE Trans. Electron. Comput. EC-10, 260-268 (1961).

[Fre74] H. Freeman: Computer processing of line-drawing images. Comput. Surv. 6, 57-97 (1974).

[Kov84] V.A. Kovalevsky: Discrete topology and contour definition. Pattern Recognition Letters 2, 281-288 (1984)

[Kov89] V.A. Kovalevsky: Finite topology as applied to image processing, CVGIP 46, 141-161 (1989).

[Kov90] V.A. Kovalevsky: New definition and fast recognition of digital straight segments and arcs. In: Proceeding of the International Conference on Pattern Recognition (ICPR-10), Los Alamitos, IEEE Computer Society Press 1990 (vol. II, pp. 31-34).

[Kov94a] V.A. Kovalevsky: Topological foundations of shape analysis. In: Ying-Lie O, Alexander Toet, David Foster, Henk J.A.M. Heijmans, Peter Meer (eds.): Shape in Picture. Berlin Heidelberg New York Tokyo: Springer 1994 (Series F: Computer and Systems Sciences Vol. 126, pp. 21-36).

[Kov94b] V.A. Kovalevsky: A new concept for digital geometry. In: Ying-Lie O, Alexander Toet, David Foster, Henk J.A.M. Heijmans, Peter Meer (eds.): Shape in Picture. Berlin Heidelberg New York Tokyo: Springer 1994 (Series F: Computer and Systems Sciences Vol. 126, pp. 37-51).

[RevDeb94] J.P. Reveilles, I. Debled-Renesson: A linear algorithm for segmentation of discrete curves. Third International Workshop on Parallel Image Analysis: Theory and Applications, Washington, 1994.

[RosKaak82] A. Rosenfeld and A. C. Kaak: Digital Picture Processing. 2nd edition, Academic Press 1982.

[Whit83] T. Whitted: Antialiased line drawing using brush extrusion. SIGGRAPH 83, 151-156 (1983).

Maximal Superpositions of Grids and an Application

Gilles d'Andréa and Christophe Fiorio

LIRMM, 161 rue Ada, 34392 Montpellier Cedex 5

Abstract. In this paper, we present an efficient algorithm of maximals superpositions of two digital sets coded in grids. It represents the basis of an application which controls the printing quality of industrial products labels, in which we consider a grid as the representation of a character.

1 Introduction

The problem we would like to solve is to control the printing quality of information on labels of industrial products. This application is different from the classical character recognition applications. Indeed, the problem here is not to identify a text, but to compare a printed text with a pattern of the text to be printed. It concerns more a default detection application than a character recognition application.

In [dF97] we have presented an algorithm which determines maximal superpositions of two horizontally convex polyominoes. We call a polyomino, a connected finite set of adjacent dashed cells lying two by two along a side in a rectangular grid. Polyominoes were first introduced by Golomb in 1954 in a talk to the Harvard Mathematics Club, see [Gol54,Gol65]. It is clear that a polyomino can represent a character, but unfortunately all characters can't be coded with a polyomino. Thus, we will present a new algorithm based on the previous one which, given two grids P and P', determines the (horizontal and/or vertical) translation of P on P' such that the cardinal of the superposed dashed cells is maximal.

The application of printing quality control is based on this algorithm and prepare the comparison between printed characters to patterns to print.

2 Superposition of two Grids

Given P et P' two digital sets coded in two grids of n (resp. n') lines, and of m (resp. m') columns, given a vertical translation, there is at most

$(m+m'-1)$ possible horizontal translations of P on P' such that the two grids have at less one cell superposed. In the same way, given a horizontal translation, there is at most $(n+n'-1)$ possible vertical translations of P on P' such as P and P' have at less one superposed cell.

Thus, we can build a matrix $M_{P/P'}$ of $(m+m'-1) \times (n+n'-1)$ cells, where each element $M_{P/P'}[h,v]$ of the matrix gives the number of superposed dashed cells for this translation of P on P' of h cells horizontally and of v cells vertically[1]. Besides, for every translation, the cardinal of the superposed dashed cells can be computed in scanning the cells set of P. The complexity of this naive algorithm is therefore in $O\left((n+n') \times (m+m') \times (n \times m)\right)$.

3 Presentation of our Algorithm

Every grid is decomposed in horizontal lines segments. The algorithm is based on this decomposition, each superposition of segments is independently processed to compute the total of superposed dashed cells.

3.1 Grids Superposing and Coding

We use a horizontal line by line coding for the grids. For every horizontal line, of the grid, every set of consecutive dashed cells (segment) is coded by a pair marking the position of the first dashed cell and the last dashed cell (see figure 1). Thus, the number of pairs coding a line is equal to the number of segments composing this line.

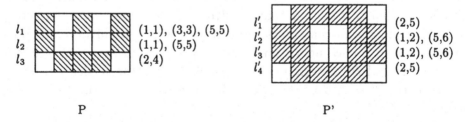

$$
\begin{array}{ll}
l_1 & (1,1), (3,3), (5,5) \\
l_2 & (1,1), (5,5) \\
l_3 & (2,4)
\end{array}
\qquad
\begin{array}{ll}
l'_1 & (2,5) \\
l'_2 & (1,2), (5,6) \\
l'_3 & (1,2), (5,6) \\
l'_4 & (2,5)
\end{array}
$$

P $\qquad\qquad$ P'

Fig. 1. Example of 2 grids and their associated coding.

Given P and P' two grids respectively coded in two sets of lines $L = \{l_1, l_2, ..., l_n\}$ and $L' = \{l'_1, l'_2, ..., l'_{n'}\}$, given v a vertical translation and h a

---

[1] h (resp. v) is positive if P' is translated to the right (resp. to the down), negative if P' is translated to the left (resp. to the up).

horizontal translation of P on P', we define $S_{P/P'}$ the function computing the number of dashed cells of P superposed to the dashed cells of P' (*overlapping surface*). A superposition of P on P' is said maximal if the overlapping surface is maximal.

Given a vertical translation v, a lines subset of P is superposed to a lines subset of P'. Given a horizontal translation h, let's define $S_{l/l'}$ the function computing the overlapping surface of two lines of the grids. Let L_v the set of lines pairs superposed of P on P' for a vertical translation v, we get:

$$M_{P/P'}[h, v] = \sum_{(l,l') \in L_v} S_{l/l'}(h)$$

Given a translation, the overlapping surface of two grids is brought back to the sum of overlapping surfaces of the lines of P superposed to the lines of P'. We are going now to discuss about the superposition of two lines.

3.2 Superposition of two Lines of two Grids

Given l and l' two grids lines, respectively defined by two sets of pairs $\{(x_1, y_1), (x_2, y_2), ..., (x_q, y_q)\}$ and $\{(x'_1, y'_1), (x'_2, y'_2), ..., (x_{q'}, y_{q'})\}$, for all the horizontal translations, every segment of l is superposed to every segment of l'. We are lead to the study of the superpositions properties of a segment of l on a segment of l'.

Given t and t' two segments of two grids lines respectively defined by pairs (x, y) and (x', y'), let's defines the function $S_{t/t'}$ that computes the overlapping surface of t on t' for a horizontal translation h.

Dashed cells of t and t' are consecutive, so all the maximal superpositions of t on t' are also consecutive and may be coded by a triplet $(B_{t/t'}, E_{t/t'}, N_{t/t'})$ (see Figure 2). With $N_{t/t'}$ the number of superposed

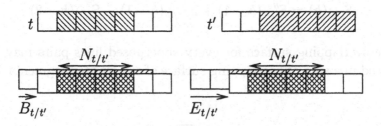

Fig. 2. maximal superpositions of 2 segments of 2 grids lines.

dashed cells, and $B_{t/t'}$ (resp. $E_{t/t'}$) the minimum shift (resp. maximum) of t giving $N_{t/t'}$, we have:

$$B_{t/t'} = min(x' - x, y' - y) \tag{1}$$
$$E_{t/t'} = max(x' - x, y' - y) \tag{2}$$
$$N_{t/t'} = min(y - x + 1, y' - x' + 1) \tag{3}$$

The functions $S'_{t/t'}$, $S''_{t/t'}$ are defined by:

$$S'_{t/t'}(h) = S_{t/t'}(h + 1) - S_{t/t'}(h) \tag{4}$$
$$S''_{t/t'}(h) = S'_{t/t'}(h) - S'_{t/t'}(h - 1) \tag{5}$$

Let's studying these functions:

h	...$B-N-1$	$B-N$	$B-N+1$	...	$B-1$	B	...	E	$E+1$	...	$E+N-1$	$E+N$	$E+N+1$ ...
$S(h)$	0	0	1	↗	$N-1$	N	—	N	$N-1$	↘	1	0	0
$S'(h)$	0	1	1	—	1	0 0	-1	-1	-1	—	-1	0	0
$S''(h)$	0	1	0	—	0	-1 0	-1	0	0	—	0	1	0

We can notice that four key-values define the function $S''_{t/t'}$. Indeed, $S''_{t/t'}(h) \neq 0$ for $h \in \{B - N, B, E, E + N\}$. So, the function $S''_{t/t'}$ can be coded by these four values, that are directly deductible of maximal superpositions triplets. Then, we can compute the function $S_{t/t'}$ from $S''_{t/t'}$ with the adequate replacements:

$$S''_{t/t'}(h) = S_{t/t'}(h + 1) - 2S_{t/t'}(h) + S_{t/t'}(h - 1) \tag{6}$$
$$S_{t/t'}(h) = S''_{t/t'}(h - 1) + 2S_{t/t'}(h - 1) - S_{t/t'}(h - 2) \tag{7}$$

The overlapping surface for every superposed lines pairs may be decomposed in a sum of overlapping surface of superposed segments. So, we get:

$$S_{l/l'}(h) = \sum_{(t,t') \in l \times l'} S_{t/t'}(h)$$

3.3 Discrete Function of two Grids Superposition

For every vertical translation v, we determine the overlapping surface of the two superposed grids by computing the sum of the function of superposed lines overlapping surfaces, so:

$$M_{P/P'}[h,v] = \sum_{(l,l')\in L_v} S_{l/l'}(h) = \sum_{(l,l')\in L_v} \sum_{(t,t')\in l \times l'} S_{t/t'}(h)$$

To avoid the calculation of the function $S_{t/t'}$ for every horizontal translation h, we use the function $S''_{t/t'}$ that is coded on four key-values. By means of the following equations, all the overlapping surfaces may be iteratively computed in a single scan.

$$M_{P/P'}[h,v] = \sum_{(l,l')\in L_v} \sum_{(t,t')\in l \times l'} \left(S''_{t/t'}(h-1) + 2S_{t/t'}(h-1) - S_{t/t'}(h-2) \right)$$

$$M_{P/P'}[h,v] = \sum_{(l,l')\in L_v} \sum_{(t,t')\in l \times l'} \left(S''_{t/t'}(h-1) \right) + \sum_{(l,l')\in L_v} \sum_{(t,t')\in l \times l'} \left(2S_{t/t'}(h-1) \right)$$

$$- \sum_{(l,l')\in L_v} \sum_{(t,t')\in l \times l'} \left(S_{t/t'}(h-2) \right)$$

$$M_{P/P'}[h,v] = \sum_{(l,l')\in L_v} \sum_{(t,t')\in l \times l'} \left(S''_{t/t'}(h-1) \right) + 2M_{P/P'}[h-1,v] - M_{P/P'}[h-2,v]$$

3.4 The Algorithm

The algorithm computes a matrix $M''_{P/P'}$, sum of functions $S''_{t/t'}$ for every pair of segments superposed. Then, the superpositions matrix $M_{P/P'}$ is computed with $M''_{P/P'}$ and the equation previously given.

3.5 Analysis of Complexity

Given P and P' two grids of n (resp. n') lines and m (resp. m') columns and given q (resp. q') the maximal number of segments in a line of P (resp. P'). The grids P and P' may be extended with empty lines and/or empty columns to N lines and M columns such that $M = max(m,m')$ and $N = max(n,n')$ without modification on q and q'.

The Algorithm 1 we can distinguish two parts: the building of the matrix $M''_{P/P'}$ and the calculation of $M_{P/P'}$ from $M''_{P/P'}$. The building of

---

Algorithm 1: Computing of the overlapping surfaces matrix $M_{P/P'}$.

    Data   : Two Grids P and P' of respectively $n \times m$ and $n' \times m'$ cells

    Result : $M_{P/P'}$ translations matrix of P on P' of $(n + n' - 1) \times (m + m' - 1)$ elements

1   *Initialization to 0 of the $M''_{P/P'}$ elements*

2   **foreach** *vertical translation v* **do**

3   **foreach** *pair (l, l') of superposed lines* **do**

4   **foreach** *segment (x, y) of line l* **do**

5   **foreach** *segment (x', y') of line l'* **do**

6 $B \leftarrow min(x' - x, y' - y)$

7 $E \leftarrow max(x' - x, y' - y)$

8 $N \leftarrow min(y - x + 1, y' - x' + 1)$

9 $M''_{P/P'}[B - N, v] + +$

10 $M''_{P/P'}[B, v] - -$

11 $M''_{P/P'}[E, v] - -$

12 $M''_{P/P'}[E + N, v] + +$

13 **for** $v \leftarrow -n + 1$ *to* $n' - 1$ **do**

14 $M_{P/P'}[-m + 1, v] \leftarrow M''[-m, v]$

15 $M_{P/P'}[-m + 2, v] \leftarrow M''[-m + 1, v] + 2M_{P/P'}[-m + 1, v]$

16   **for** $h \leftarrow -m + 3$ *to* $m' - 1$ **do**

17 $M_{P/P'}[h, v] \leftarrow M''[h - 1, v] + 2M_{P/P'}[h - 1, v] - M_{P/P'}[h - 2, v]$

---

the matrix $M''_{P/P'}$ (line 1 to 12) is in $O(N^2 \times (q \times q'))$. The computing of the matrix $M_{P/P'}$ (line 13 to 17) requires $O(M \times N)$ operations. So, the global complexity of the algorithm is in $O(N \times (M + N \times (q \times q')))$

In considering that the maximal number of segments in a line is bounded, the complexity becomes $O(N \times (M + N))$. Such hypothesis is reasonable in a concrete application, like the research of maximal superpositions of two characters. In the worst case where the number of segments is in $O(M)$ (a checkerboard grid for example), we achieve an equivalent complexity of the naive algorithm discussed in section 2.

4   A Concrete Application

In the food or pharmaceutical industry, the impression of expiration dates or informations on products must remain perfectly readable. Printing processes often generate defaults. The goal of our application, is to verify if the printed text is sufficient in quality and without ambiguity.

In our application, every character is coded by a matrix such that for every dashed cell, an ink-bubble will be projected for the printing. The assessment of the printing quality consists therefore to compare the *characters to print* to *printed characters*.

4.1 Characters Segmentations and Deformations Analysis

Fig. 3. Characters to Print

Fig. 4. Printed Characters

The printing matrix of every character is coded by a digital set where each projected ink bubble corresponds to a dashed cell (see figure 3). The printed characters undergo the distortions owed to the printing process (trajectories of inks bubbles are not perfect) and the picture aquisition process (the picture resolution is such that every ink bubble covers a set of dashed cells, see figure 4).

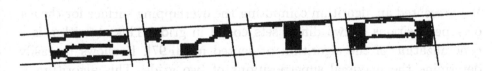

Fig. 5. Segmented Printed Characters

The printed characters undergo distortions of scaling, rotations and slanting that are estimated by the analysis of the parallelogram including the printed text. A space analysis between characters allows to separate every character (see [CL96] and Figure 5). Then, every character is coded as a digital set (grid).

A scaling, rotation, and slant similar to the printed characters is applied to the character to print in order to get two sets of characters theoretically identical, in size, rotations and slants. Differences between grids of the printed characters and grids of the characters to print after distortions are generated by the imprecisions of inks bubbles trajectories. The analysis of these differences is the basis to decide if a character still remains readable without ambiguity.

4.2 Characters Superpositions

For every character, we are looking for one maximal superposition of the grid to print on the printed polyomino to determine the differences. In the characters case, the maximal number of segments in a horizontal line is bounded to four with the character "M" and "W", so the algorithm presented here, is very convenient.

A first method to compare the two superposed grids consists in counting all dashed cells which are exclusively in one set. This brutal method has for inconvenience to not hold on with the characters geometry. Indeed, deviations of projected ink bubbles, or their missing can create or break connexity between the different parts of characters. A "O" can easily become a "C", and vice versa.

The characters printing quality depends on all of these parameters: distortions owed to rotations, slants and scales, as well as the add or loss of connexity between the different parts of the character. In quantifying these parameters and in weighting them, we get a quality criterion for every printed character on which is based our application.

5 Conclusion

We presented an algorithm computing the overlapping surface for the set of superpositions of two difital sets coded in grids. This algorithm is a generalization of the algorithm presented in [dF97]. It allows to easily determine the maximal superpositions of two grids. This algorithm is used for a practical application of printing quality control where we must determine if characters have been correctly printed, i.e. every printed character can be read without ambiguity.

This algorithm is linear in the size of the superpositions matrix if the number of segments composing a line of the grids remains limited. In the general case, the complexity is equivalent to the naive algorithm presented in section 2. But in our application, where the grids to be superposed are characters, thus the number of segments of a line is bounded and we can benefit the improvement brought by our algorithm.

References

[CL96] R. G. Casey and E. Lecolinet. A survey of methods and strategies in character segmentation. *IEEE Transactions on Pattern Analysis and Machine Intelligence*, 18(7):690–706, July 1996.

[dF97] G. d'Andréa and C. Fiorio. Superpositions maximales de polyominos horizontalement convexes. Rapport de Recherche LIRMM, Laboratoire d'Informatique, de Robotique et de Microélectronique de Montpellier, 161 rue Ada – 34392 Montpellier Cedex 5 – France, July 1997.

[Gol54] S.W. Golomb. Checkerboards and polyominoes. *American Mathematical Monthly, LXI*, 10:672–682, December 1954.

[Gol65] S.W. Golomb. *Polyominoes*. Charles Scribner's Sons, New-York, 1965.

Discrete Shapes and Planes

Multiresolution Representation of Shapes in Binary Images II: Volume Images

Gunilla Borgefors[1], Gabriella Sanniti di Baja[2], and Stina Svensson[1]

[1] Centre for Image Analysis, Swedish University of Agricultural Sciences,
Lägerhyddvägen 17, SE-752 37 Uppsala, SWEDEN
email:gunilla,stina@cb.uu.se
[2] Istituto di Cibernetica, Italian National Research Concil, Via Toiano 6, IT-80072
Arco Felice (Naples), ITALY
email:gsdb@imagm.na.cnr.it

Abstract. Multiresolution representations of discrete patterns are of great interest, specially when working with volume images. The huge amount of data that volume images contain at high resolution can be considerably compressed at lower resolution, and while the obtained representation still can be suited for simple shape analysis tasks, provided that shape is adequately preserved when resolution decreases. In this paper, we present new methods for building shape preserving binary resolution pyramids in three dimensions. The performance of the methods is quantitatively evaluated.

1 Introduction

In [1], methods for building shape preserving binary resolution pyramids in two dimensions are described. In this paper we present the corresponding methods in three dimensions.

Multiresolution representations of discrete patterns are of great interest for several reasons. Whatever image application, the best resolution can be used at each step, and the results can be propagated through the pyramid. Being able to use low resolution, i.e. small images, in parts of the computations becomes especially important for volume images, as they contain huge amounts of data.

A typical example where resolution pyramids are useful is the matching phase in object recognition. Typically, the discrete pattern (as well as any available prototype) is represented by a graph, whose number of nodes depends on resolution. Graph comparison can first be performed using all prototypes in the library, but only at low resolution levels, where a small number of nodes are involved. This will sort out the most promising matchings. Detailed comparison can then be performed with a reduced number of prototypes at higher resolution levels, where a larger number of nodes have to be taken into account.

In two dimensions the most straightforward example of a multiresolution shape representation is the 2×2 binary pyramid. The corresponding multiresolution image in three dimensions is the $2 \times 2 \times 2$ binary pyramid, where voxels are either black (pattern) or white (background). Each level of the pyramid is a three dimensional array, where the size of the array is 1/8 of the size of the array at the immediately previous higher resolution level. A decimation process is used to build the successive resolution levels starting from the highest, original, resolution level. The next, lower, resolution level is built by partitioning the array in $2 \times 2 \times 2$ blocks of voxels, *children*, and associating a single voxel, *parent*, to each block. The new voxel is set to black or white depending on the colour of its children (and, in some cases, of their neighbours), according to some fixed rule. The process is repeated for the previously computed lower resolution representation, and then further iterated to build all possible resolution levels, ending in a single voxel in the lowest resolution level. If the original image is not $2^n \times 2^n \times 2^n$, planes are added as needed.

As it was the case in two dimensions, rules more sophisticated than the logical OR and AND operations should be used to build a shape preserving binary pyramid. In fact, all the shortcomings of these occurring in two dimensions are equally present in three dimensions, where shape modifications occur even more rapidly. If OR-pyramids are built, the set of black voxels soon becomes a large amorphous blob, due to the large number of $2 \times 2 \times 2$ configurations that contain at least one black voxel and thus get a black parent. Pattern regions sufficiently close to each other merge, filling in tunnels and cavities initially present in the pattern, or causing creation of new tunnels and cavities. If AND-pyramids are built, the pattern is soon dramatically shrunk. Narrow regions of the initial pattern either completely vanish or become disconnected as the resolution decreases. This is because only the $2 \times 2 \times 2$ configuration consisting of eight black voxels gets a black parent. In both cases, the shape of the pattern is not adequately preserved. Something in between the AND and OR pyramids is needed to produce better results as concerns shape preservation.

The position of the grid used to partition the array into $2 \times 2 \times 2$ blocks also affects the shape of the resulting pattern. The grid can be shifted in eight different positions over the image, originating eight differently shaped patterns at the next resolution level. A combination of the eight possible resulting images is expected to produce more stable representations. The general problem of placement of a lower resolution grid over a higher resolution becomes worse the higher the dimension of the image is, but is seldom addressed in the literature.

It should be remarked that changes of the digital topology of an object are unavoidable when image resolution is decreased. It is impossible to represent the connectivity, tunnels, and cavities of a three dimensional object unless there is a sufficient number of voxels. For this reason, we do *not* claim to preserve topology using the methods in this paper, but concentrate on perceived shape. In an attempt to quantify the goodness of our methods, we also introduce a voxel-counting measure. However, we do take topology into account when possible and it *is* fairly well preserved, especially as concerns maintenance of pattern connectedness.

2 Some Definitions

Let I_1 be the $2^n \times 2^n \times 2^n$ original black and white image, stored in the bottom level (the first level) of the pyramid. Each of the successive n levels of the pyramid $(I_2, I_3, ..., I_{n+1})$ is built from its preceding level. Every voxel of level I_k, $2 \leq k \leq n - 2$, is either black or white, depending on the colours of the voxels in I_{k-1}. Since the last three levels consist of only 1, 8, and 64 voxels, respectively, they are not meaningful for shape representation. We define, in the following, our resulting resolution pyramid as a structure consisting of $n - 2$ levels, where the $8 \times 8 \times 8$ level is the last one.

All our methods will be illustrated on the same small ($2^6 \times 2^6 \times 2^6$) test image shown in Fig. 1. The image consists of a cylinder which ends in a cone and on the top of that a sphere with a tunnel, all in the digital Euclidean metric. This $2^6 \times 2^6 \times 2^6$ image is the first level of the pyramid, I_1. As this level is always the same, whatever scheme used to build the pyramid, we will in the following show only the three lower levels, consisting of $2^5 \times 2^5 \times 2^5$, $2^4 \times 2^4 \times 2^4$, and $2^3 \times 2^3 \times 2^3$ voxels, respectively. Using a larger image as an illustration is not necessary, as shape and topology preservation problems are more prominent at low resolutions. We have, however, tested the methods on many different images. A few examples will be discussed in Section 6.

Fig. 1. First level, I_1, of a binary pyramid, where a $2^6 \times 2^6 \times 2^6$ test image is stored.

3   Combining Eight Binary Images

The first approach to construct a shape preserving 3D pyramid is based on the combination of the eight binary images, which are obtained by shifting the partition grid to all possible positions. For a given position of the partition grid, an image of the same size as the successive level of the pyramid is built from its preceding level, using a simple rule. We might resort to the commonly used OR and AND operations to decide about voxel colour, or use mathematical operations counting the number of black and white children. The latter possibility is more flexible than simply using OR and AND operations and will be preferred in this paper. The eight resulting images are then combined, again using some arithmetic rule, to form the successive level of the pyramid. This new level is then partitioned in the eight possible ways, and the next level is built, as before. The process continues until level $8 \times 8 \times 8$.

The building rules we adopt compute the sum of the values of the eight children, $\sum v$. These rules also allow us to obtain the OR- and the AND-pyramid. In the OR-pyramid a parent becomes black if $\sum v \geq 1$ and in the AND-pyramid a parent becomes black $\sum v = 8$. Other arithmetic rules could be used, based on $\sum v$, i.e. $\sum v \geq n$, where $n = 2, ..., 7$. Although the shapes of the resulting patterns are better than those provided by the OR- and the AND-pyramid, shape and topology are still barely preserved if we use only a single grid position.

Formally, the pyramid is built in the following way. Let $A_{i+1,1}$, $A_{i+1,2}$, $A_{i+1,3}$, $A_{i+1,4}$, $A_{i+1,5}$, $A_{i+1,6}$, $A_{i+1,7}$ and $A_{i+1,8}$ be the eight different images, obtained by shifting the grid used to partition I_i into $2 \times 2 \times 2$ blocks. The rule $\sum v \geq n$, where $n = 1, ..., 8$, is used to build $A_{i+1,j}$, $j = 1, ..., 8$. When combining these eight images, the sum of the eight voxels values in the same position, $\sum g$, is computed for all voxels. The combined level I_{i+1} is obtained setting to black all voxels with $\sum g \geq m$, where $m = 1, ..., 8$.

Pyramid rule and image combinations where $\sum v \geq n$ and $\sum g \geq 9 - n$, $n = 1, ..., 8$, give satisfying results, see Section 6. In Fig. 2 we show the result when $n = 1$ is used, which is equivalent to first "OR-ing" and then "AND-ing."

Note that although these "combined" pyramids are much less sensitive to the placement of the $2 \times 2 \times 2$ grid, the results are still not translation independent. When the grid is shifted a new plane of white voxels must be added. Depending on in what direction the grid is shifted, the plane can be placed at the top or at the bottom, in front or behind, or, to the left or to the right of the image, which again gives eight possibilities.

4   Using Intermediate Grey-Level Images

Handling eight volume images simultaneously is per definition time and memory consuming. Therefore, we have devised a method where a combination of the images that would be obtained in correspondence with the different grid positions can be achieved without actually building the eight images. The combination is coded as an intermediate grey-level image, which contains the same information

79

Fig. 2. Pyramid computed by AND-ing eight OR images

as the eight shifted binary images. By binarizing this grey-level image, according to some rule, the next binary pyramid level can be computed.

Suppose that v is a voxel in level i, having even coordinates, i.e. v belongs to row $2j$, column $2k$ and plane $2l$, for some j, k and l. Depending on the position of the grid on the i level image, the voxel v belongs to one of the eight $2 \times 2 \times 2$ blocks shown in Fig. 3. Each of these blocks would generate one voxel (either black or white) on each of the eight next level binary images. The grey-level image at resolution level $i + 1$, G_{i+1}, of the pyramid is built by taking into account the value that v would have in all eight images, by using the $3 \times 3 \times 3$ linear filter shown in Fig. 4. The weights in the filter correspond to the number of times a neighbouring voxel to v is present in the blocks in Fig. 3. The voxel itself is present in all, whereas the point-neighbours are present only in a single block. In G_{i+1} the possible values of a voxel v ranges from 0 (when v and all its neighbours are white) to 64 (when v and all its neighbours are black).

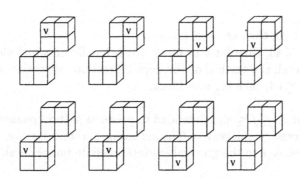

Fig. 3. Possible positions of the grid around the voxel v.

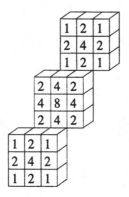

Fig. 4. Mask of weights for computing the intermediate grey-valued image.

The intermediate image G_{i+1} can be binarized by using different criteria. The simplest would be to threshold the image at a suitable level, but we achieved better results by more sophisticated rules. Note that, simple thresholding by setting all voxels with grey-level $f > 0$ ($f = 64$) to black does *not* produce the OR (AND) image.Generally, voxels with small values (close to 0) should become white, as they have few black neighbours and would probably be white in the next level of the binary pyramid in any grid position. Analogously, voxels with large values (close to 64) should be set to black, as they have mostly black neighbours and would probably be black in almost all eight images obtained by shifting the grid. For the voxels with intermediate values, the final colour is decided not only by the value of the voxel itself, but also by the values of its neighbours. This means that the values of a parent is based not only on the values of its children, but on the values in a $7 \times 7 \times 7$ neighbourhood.

The two best criteria for shape preservation that we have found are the following. Let a voxel v have value f.

- **Criterion 1:**
   If $0 \leq f \leq 20$, then set v to white.
   If $21 \leq f \leq 43$, then set v to black if it is locally maximal along either the horizontal, the vertical or the depth direction, otherwise set it to white.
   If $44 \leq f \leq 64$, then set v to black.

The rule for $21 \leq f \leq 43$ is intended to preserve pattern connectedness while avoiding unnecessary thickening. The resulting pyramid for the test image by applying *Criterion 1* to the grey-valued intermediate images is shown in Fig. 5.

- **Criterion 2:**
   If $0 \leq f \leq 20$, then set v to white.
   If $21 \leq f \leq 43$, then set v to black if f is larger than the average value of a $3 \times 3 \times 3$ neighbourhood centred on v, otherwise set it to white.
   If $44 \leq f \leq 64$, then set v to black.

The rule for $21 \leq f \leq 43$ is intended to preserve the most "significant" out of several neighbouring voxels. The resulting pyramid for the test image by applying *Criterion 2* to the grey-valued intermediate images is for our simple example *exactly* the same as for *Criterion 1*, so see Fig. 5 again, but this is not usually the case.

Fig. 5. Pyramid made by using intermediate grey-valued image levels using *Criterion 1*. The tunnel in the sphere is preserved in all three resolution levels.

5 Implementations

The implementations of the described algorithms are written in C and integrated in IMP (IMage Processing), a general image analysis software developed at *Centre for Image Analysis*. The software is using the X-Windows standard and Motif for the user interface. For the visualization of the images we also use IMP.

Even though we are working with 3D images, when run on a DEC Alpha (a standard UNIX workstation) it takes about one second to compute the binary pyramid for a $128 \times 128 \times 128$ image when using the method with intermediate grey-level images. The pyramid that combine eight binary images at each level is slower to build, but the times are still reasonable. It takes about 5 seconds for a $64 \times 64 \times 64$ image and about 40 seconds for a $128 \times 128 \times 128$ image.

6 Results

The results for our methods and the motivations for the thresholds at *Criterion 1* and *2* are presented in this section, using three different test images. The first one is the same as earlier, see Fig. 1. To build the second image we have

Fig. 6. Test image consisting of six hollow spheres connected by three cylinders

Fig. 7. Test image consisting of an HIV particle

connected six hollow spheres by three cylinders, see Fig. 6. The third is a real image consisting of an HIV particle, see Fig. 7.

To get one objective, quantitative measure of the performance of our methods, we have counted object voxels in the different pyramid levels for the different methods. Let $V(I_i)$ be equal to the number of object voxels in level I_i. We define the measure $M(I_i)$ as

$$M(I_i) = \frac{8^{i-1} \times V(I_i)}{V(I_1)},$$

i.e. when the $M(I_i)$ is close to 1.00 the object in level I_i contains the "correct" number of object voxels. If it is > 1 there are "too many" and if < 1 "too few". This gives some sort of measure of how good the methods are even if it does not take shape into account. Since *Criterion 1* and *Criterion 2* give very similar results, we only present both methods for the first image. We have also looked at the topology of the results for a large number of test images. The conclusions support those for the presented examples.

- **Image one, the tower in Fig. 1** The original image contains 20035 voxels. In Table 1 the result is shown for the method of combining eight binary images, and in Table 2 the result when using the intermediate grey-level image. When using the method of combining eight binary images at each level according to the arithmetic rules $\sum v \geq n$ and $\sum g \geq 9 - n$, the connectivity is preserved for $n = 1, ..., 6$. The grey-level method works well for all the thresholds in the image, from the connectivity point of view.

- **Image two, the spheres in Fig. 6** The original image contains 23172 voxels. In Table 3 the result is given when using the intermediate grey-level image. A wide range of thresholds give similar results. In the $8 \times 8 \times 8$ image the cavities in the spheres are not preserved and when the lower threshold is low the six spheres turns into a blob-looking thing.

- **Image three, the HIV particle in Fig. 7** The original image contains 13396 voxels. See Table 4 for the result when using the intermediate grey-level image. The shape of the particle is well preserved in level I_2 and I_3. Because of the complex structure the object is hard to represent in an $8 \times 8 \times 8$ image.

Table 1. The result for image one when building the pyramid by using a combination of binary images.

$\sum v$	$\sum g$	$M(I_2)$	$M(I_3)$	$M(I_4)$
1	8	1.01	0.98	0.95
2	7	1.01	0.97	0.74
3	6	1.00	0.98	0.82
4	5	1.00	0.97	0.61
5	4	0.99	0.96	0.69
6	3	1.00	0.95	0.61
7	2	0.99	0.97	0.69
8	1	1.00	0.97	0.69

Table 2. The result for image one when building the pyramid by using intermediate grey-level images. The first three columns describe the result when using *Criterion 1* and the last three when using *Criterion 2*.

lower threshold	higher threshold	$M(I_2)$	$M(I_3)$	$M(I_4)$	$M(I_2)$	$M(I_3)$	$M(I_4)$
16	44	1.03	1.07	1.23	1.03	1.07	1.23
17	44	1.03	1.06	1.20	1.03	1.06	1.20
18	44	1.01	1.03	1.12	1.01	1.03	1.12
19	43	1.01	1.03	1.15	1.01	1.03	1.15
19	44	1.01	1.02	1.12	1.01	1.02	1.12
19	45	1.01	1.02	1.12	1.01	1.02	1.12
20	43	1.01	1.01	1.10	1.01	1.01	1.10
20	44	1.01	1.00	1.07	1.01	1.00	1.07
20	45	1.01	1.00	1.07	1.01	1.00	1.07
21	43	1.01	1.00	1.07	1.01	1.00	1.07
21	44	1.00	1.00	1.07	1.00	1.00	1.07
21	45	1.00	1.00	1.07	1.00	1.00	1.07
22	44	0.97	0.97	0.89	0.97	0.97	0.89

Table 3. The result for image two when building the pyramid by using intermediate grey-level images and binarize the images with *Criterion 1*.

lower threshold	higher threshold	$M(I_2)$	$M(I_3)$	$M(I_4)$
16	44	0.98	1.01	1.81
17	44	0.98	1.00	1.75
18	44	0.98	0.99	1.59
19	43	0.96	0.98	1.37
19	44	0.96	0.97	1.35
19	45	0.95	0.96	1.28
20	43	0.96	0.95	1.22
20	44	0.96	0.94	1.17
20	45	0.95	0.93	1.10
21	43	0.94	0.93	1.08
21	44	0.94	0.93	1.06
21	45	0.93	0.92	0.99
22	44	0.94	0.89	0.93

From this, we can conclude that values of the lower threshold 20 and the higher threshold 44 in the method that uses intermediate grey-level images are not very critical. A small change of threshold does not change the results significantly. Connectivity, tunnels, and cavities are preserved in all images except for some cases of the $8 \times 8 \times 8$ image.

When we use the method that combine eight binary images at each level, $\sum v \geq n$ and $\sum g \geq 9 - n$, where $n = 1, ..., 8$, the results are quite similar for all n, and all are significantly better than using just the OR or the AND operations. It should, however, be pointed out that the best method to preserve

Table 4. The result for image three when building the pyramid by using intermediate grey-level images and binarize the images with *Criterion 1*.

lower threshold	higher threshold	$M(I_2)$	$M(I_3)$	$M(I_4)$
16	44	1.00	1.34	2.18
17	44	1.00	1.31	1.87
18	44	0.99	1.26	1.57
19	43	0.99	1.19	1.41
19	44	0.98	1.19	1.41
19	45	0.97	1.19	1.41
20	43	0.96	1.11	1.07
20	44	0.95	1.11	1.07
20	45	0.94	1.11	1.07
21	43	0.95	1.07	1.07
21	44	0.94	1.07	1.07
21	45	0.93	1.07	1.07
22	44	0.92	1.01	0.96

thin links is to AND eight OR-images. This result differs from the 2D case, where intermediate values of v and g were found to be preferable.

7 Conclusions

Several new methods have been presented that improve shape preservation when representing volumetric patterns in binary resolution pyramids. Our first approach uses only binary images. A binary pyramid is built by using suitable combinations of the eight images obtained by shifting the $2 \times 2 \times 2$ partition grid. The rules are simple: first count the number of black children and build the eight possible lower resolution images, then count the number of times a voxel is black in any of the eight images. We have found that AND-ing eight OR-images generally gave the best results for the many patterns we have tested.

Our second approach replaces the eight binary images with one grey-level image, computed by a $3 \times 3 \times 3$ linear filter, which take into account the eight $2 \times 2 \times 2$ configurations that can be placed within the $3 \times 3 \times 3$ neighbourhood of a voxel. This grey-valued image is binarized by double thresholding and any of two proposed intermediate rules. These rules do not only consider the grey-level value of the voxel itself, but also those of its neighbours, thus using the context of a voxel to determine its significance in the pattern. This method is much faster than the previous one.

To determine the necessary thresholds in the grey-level methods, we have introduced a simple quantitative measure of the quality of the results. It is shown that a small change in threshold values does not change the resulting pyramid very much.

Our methods are easy to implement and produce much better results than the ones obtained by the "standard" OR/AND pyramids. They also preserve topology reasonably well, but topology can not be accurately preserved in decreasing resolution no matter what you do. This is especially true in three dimensions.

Acknowledgements

The authors wish to thank Giuliana Ramella for the implementation in the 2D case. Also, thanks to Ingela Nyström, for great support with the implementation in the 3D case.

References

1. G. Borgefors, G. Ramella, and G. Sanniti di Baja. Multiresolution representation of shape in binary images. In S. Miguet, A. Montanvert, and S. Ubéda, editors, *Discrete Geometry for Computer Imagery (DGCI'96)*, pages 51–58. Springer Verlag, Berlin 1996. Lecture Notes in Computer Science 1176.

Coplanar Tricubes

Jean-Maurice Schramm

Laboratoire des Sciences de l'Image, d'Informatique et de Télédétection
Université Louis Pasteur, 7, rue René Descartes, 67084 Strasbourg cedex, France
e-mail : jms@dpt-info.u-strasbg.fr

Abstract.

Within the framework of the arithmetic discrete geometry introduced by J.P.Reveillès, I. Debled has defined the concept of tricubes and found out that the total number of the tricubes that may appear in a naive plane is fourty.

This study concerns the *coexistence* of tricubes in a plane. We call *complete combination of coplanar tricubes* any set of tricubes, such as a naive plane exists, which contains all the tricubes of this set without any other. We present an algorithm which calculates the set of these combinations.

It appears that the number of these combinations is quite small : only 99, although a "combinatory explosion" could have been expected during their calculation. Their list is given in the appendix.

Key Words : arithmetic discrete geometry, discrete planes, Fourier's algorithm.

1. Introduction.

This work has been conducted in the frame of the theory of the arithmetic planes introduced by J.P.Reveillès [Rev:91], [Deb,Rev:94]. These are defined by means of two inequations. I.Debled has worked on the problem of the recognition of these planes [Deb:95]. She has introduced a local concept: the tricubes. The tricubes can be seen as little components of discrete plane, their projection on the horizontal plane being a square 3×3 : they were fourty tricubes [1]. In another paper, J.P.Reveillès has

---

[1] This result is valid for the planes which have a certain width, called naive planes, and save on symmetries. (See §1.2). Other work [Fra, Sch, Taj: 96], using Fourier's algorithm, has led the same results.

shown that each given plane could not contain more than nine different tricubes [Rev:95].

We study the compatibility between the tricubes. We call *complete combination of coplanar tricubes* any set of tricubes, such as a naive plane exists, which contains all the tricubes of this set without any other. We present an algorithm which calculates the set of these combinations. In order to do this, sets of planes are associated with the different combinations of tricubes. Each one is specified by a system of linear constraints, relating to its coefficients. This approach allows to use the Fourier's algorithm.

1.1 Outline.

The first part of this paper defines precisely the context : we call back the definition of naive planes and tricubes (1.2). Then we define relations and functions which connect the tricubes to the planes, and we introduce the concept of *complete combination of multi-coplanar tricubes* which allows a simpler calculation (1.3). These definitions allow to state precisely the treated problem (1.4).

The second part contains first a study of the properties of the concepts which have been introduced (2.1). Then the properly so called calculation is approached, first the calculation of the *complete combinations of multi-coplanar tricubes* (2.2), then we infer the *complete combinations of coplanar tricubes* (2.3).

A short conclusion (3) is followed by references and results in the appendix.

1.2 Recalls on naive planes and tricubes.

The set of naive planes $\mathbb{P}$.

The discrete planes used in arithmetic discrete geometry (see [Rev:91], [Deb, Rev:94]), are defined as sets of points (x, y, z) of $\mathbb{Z}^3$ satisfiying two inequations $0 \leqslant ax + by + cz + d < \omega$, where all parameters are integers, a, b, c are not all null and $\omega > 0$. The discrete plane thus defined is called *naive* if $\omega = sup(|a|, |b|, |c|)$. The triplet (a, b, c) is called *normal* of the plane.

Symmetry reasons allow to limit the study to the planes defined by inequations which parameters satisfy $0 \leqslant a \leqslant b \leqslant c$, that is 1 case among 48 . Otherwise :

Proposition 1.
Two naive planes which admit the same normal can be deduced one from the other by a translation from $\mathbb{Z}^3$.
Proof : For the detailled proof of the stated propositions see [Sch:97].

Therefore, in this work, we consider only the discrete naive planes defined by inequations as above, but with $d = 0$. All the naive discrete planes can be reduced to these by symmetries and translations. We note $\mathbb{P}$ the set of the normal corresponding to these planes. Thus, $\mathbb{P} = \{(a, b, c) \in \mathbb{Z}^3 \mid 0 \leqslant a \leqslant b \leqslant c , \ 0 < c \}$, but we will often interpret the elements of $\mathbb{P}$ as the naive discrete plane they define.

Representation of some sets of planes by linear constraints systems.
The sets of planes $P \in \mathcal{P}(\mathbb{P})$ [2] may be infinite and so one cannot represent them in extension, that is, element by element. On the other hand, it is possible to define some of these sets by a typical property.

We consider *linear constraints systems* : they are expressions built from linear constraints by usual logical operators $\vee$, $\wedge$, $\neg$ [3]. The *linear constraints* get on the coefficients a, b, c which are used to specify the planes and have the form $\alpha\, a + \beta\, b + \gamma\, c \succ \delta$, where $\alpha, \beta, \gamma, \delta \in \mathbb{Z}$ and $\succ$ is one of the comparison operators $>$, $\geq, <, \leq$. Such a constraint is called *homogeneous* if $\delta = 0$.

When a linear constraints system is reduced to a *conjunction* of constraints, it is possible to use Fourier's algorithm (see [Fra,Sch,Taj:96]) to decide if the system admits a solution.

In the general case, the system can be transformed into an equivalent system in the form of a disjunction of constraints conjunctions [4]. Fourier's algorithm may then be used to decide the existence of solutions, but it has to be applied to each of the conjunctions of the disjunction.

This method allows particulary to test if :
- a constraint s is a consequence of a system S : $\quad S \Rightarrow s \quad$?
- a system S_1 implies a system S_2 $\qquad$: $\quad S_1 \Rightarrow S_2 \quad$?
- two systems S_1, S_2 are equivalent $\qquad$: $\quad S_1 \Leftrightarrow S_2 \quad$?

We note
- $\mathbb{S}$ the set of these systems,
- $\mathbb{C}$ the set of conjunctions of linear constraints (we consider that $\mathbb{C}$ is a subset of $\mathbb{S}$)
- *sol* the function from $\mathbb{S}$ to $\mathcal{P}(\mathbb{P})$ which associates with each system of linear constraints the set of planes which satisfy this system.

The set of tricubes $\mathbb{T}$.
We note $\mathbb{T}$ the set of the functions t from $\{-1,0,+1\}^2$ to $\mathbb{Z}$ such as $t(0,0)=0$; these functions are called (*potential*) *tricubes*.

1.3 Relations between the combinations [5] of tricubes and the sets of planes.
The naive discrete planes whose normal (a, b, c) satisfies the conditions $0 \leq a \leq b \leq c$ are functional in x, y. For each $p \in \mathbb{P}$, for each $x,y \in \mathbb{Z}$, we note p' the so defined function.

---

[2] $\mathcal{P}(E)$ represents the set of all subsets of E.
[3] $\vee, \wedge, \neg$: *or, and, not* .
[4] The constraints are those of the given system or their negation.
[5] In this work, the word "combination" is synonymous with "set".

Let ψ be the function from $\mathbb{P} \times \mathbb{Z}^2$ to $\mathbb{T}$ defined by
$$\psi(p, (x, y))(\alpha, \beta) = p'(x+\alpha, y+\beta) - p'(x, y) \quad , \qquad (\alpha, \beta \in \{-1, 0, +1\}) .$$
This function associates with each couple formed by a naive plane p and a point (x, y) from the plane $\mathbb{Z}^2$, the tricube appearing in this plane at the point whose projection on the plane $z=0$ is the point (x, y).

Let ρ be the relation from $\mathbb{P}$ to $\mathbb{T}$ defined by
$$\rho(p, t) \Leftrightarrow \exists (x, y) \in \mathbb{Z}^2 , \psi(p, x, y) = t .$$
When $\rho(p, t)$ is true, one can note that p *contains* t, or that t *appears in* p.

Example : The following figure represents a part of the naive plane defined by the inequalities $0 \leqslant 3x + 5y + 7z < 7$ and some tricubes which appear in this plane. (The points of the plane are represented by cubes. The numbers of the tricubes are these given in the appendix A). This plane contains the tricubes 7, 8, 19, 20, 25, 37, and 40.

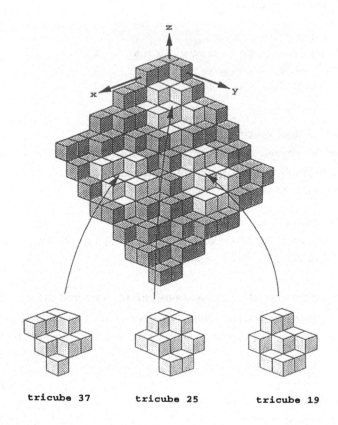

tricube 37   tricube 25   tricube 19

As it is possible for each relation, we can define the function φ associated with ρ :

$$\varphi : \ \mathbb{P} \to \mathcal{P}(\mathbb{T})$$

$$p \dashrightarrow \varphi(p) = \{t \in \mathbb{T} \mid \rho(p, t)\} = \psi(p, \mathbb{Z}^2) .$$

The function φ associates with each plane, the combination of the tricubes which appear in this plane .

We define a bijection which associates some tricubes combinations with a set of naive planes. This will help us to calculate with planes rather than directly with tricubes.

First, let γ be the function associated with the relation ρ^{-1}

$$\gamma : \ \mathbb{T} \to \mathcal{P}(\mathbb{P})$$

$$t \dashrightarrow \gamma(t) = \{p \in \mathbb{P} \mid \rho(p, t)\} .$$

The function γ associates with each tricube the set of planes which contain it.

We extend φ into the function ϕ :

$$\phi : \ \mathcal{P}(\mathbb{P}) \to \mathcal{P}(\mathbb{T})$$

$$P \dashrightarrow \phi(P) = \cap_{p \in P} \varphi(p) .$$

The function ϕ associates with each set of planes the combination of tricubes which appears in all the planes of this set. (But these planes may contain other tricubes)

In the same way we extend γ into the function Γ :

$$\Gamma : \ \mathcal{P}(\mathbb{T}) \to \mathcal{P}(\mathbb{P})$$

$$T \dashrightarrow \Gamma(T) = \cap_{t \in T} \gamma(t) .$$

The function Γ associates with each combination of tricubes the set of the planes which contain all the tricubes of the given combination (but these planes may contain other tricubes).

Eventually, we define the function Δ :

$$\Delta : \ \mathcal{P}(\mathbb{T}) \to \mathcal{P}(\mathbb{P})$$

$$T \dashrightarrow \Delta(T) = \{p \in \mathbb{P} \mid \varphi(p) = T\} .$$

The function Δ associates with each combination of tricubes the set of the planes which contain all the tricubes of the given combination and no other tricube.

We simplify the calculation by introducing the following definition :
A *complete combination of multi-coplanar tricubes* is a set T of tricubes such as it exists a *set* of planes for which T is the set of the tricubes appearing in all these planes.

1.4 The problem.

The objective of this work is to calculate the *set of the complete combinations of coplanar tricubes*, that is to say the set of the combinations of tricubes such as a naive plane exists, which contains all the tricubes of this combination, without any other. This set is $\varphi(\mathbb{P}) = \Delta^{-1}(\mathcal{P}(\mathbb{P}) - \{\varnothing\})$.

The calculation of Δ appeals to unions and differences (see proposition 7). This implies that the set of the planes of form $\Delta(T)$ is not always convex. For this reason, we calculate first the set $\phi(\mathcal{P}(\mathbb{P}))$ of complete combinations of multi-coplanar tricubes. Hence, we get $\Gamma(\mathcal{P}(\mathbb{T}))$ (see proposition 5). The elements of $\Gamma(\mathcal{P}(\mathbb{T}))$ are convex sets which can be specified by conjunctions of linear constraints.

2. Calculation.

2.1 Basic properties.
<u>Proposition 2</u> .
$\forall P_1, P_2 \in \mathcal{P}(\mathbb{P}): \phi(P_1 \cup P_2) = \phi(P_1) \cap \phi(P_2)$,
$\forall T_1, T_2 \in \mathcal{P}(\mathbb{T}): \Gamma(T_1 \cup T_2) = \Gamma(T_1) \cap \Gamma(T_2)$.
It results that ϕ and Γ are decreasing for the inclusion.

<u>Proposition 3</u>.
The function $\phi \circ \Gamma$ from $\mathcal{P}(\mathbb{T})$ to $\mathcal{P}(\mathbb{T})$, and the function $\Gamma \circ \phi$ from $\mathcal{P}(\mathbb{P})$ to $\mathcal{P}(\mathbb{P})$ are increasing for the inclusion.
They can be written
$$\phi(P) = \{t \in \mathbb{T} \mid \forall p \in P , \rho(p,t)\} ,$$
$$\Gamma(T) = \{p \in \mathbb{P} \mid \forall t \in T , \rho(p,t)\} .$$
It results that :

<u>Proposition 4</u>.
For each P in $\mathcal{P}(\mathbb{P})$, we have $\Gamma(\phi(P)) \supset P$ and,
for each T in $\mathcal{P}(\mathbb{T})$, we have $\phi(\Gamma(T)) \supset T$.

We consider the subsets of $\mathbb{P}$ which verify the property to be the set of the planes which contain (at least) a given set of tricubes. The set of the subsets of $\mathbb{P}$ which verifies such a property is the set $\Gamma(\mathcal{P}(\mathbb{T}))$.

<u>Proposition 5</u>.
The sets $\phi(\mathcal{P}(\mathbb{P}))$ and $\Gamma(\mathcal{P}(\mathbb{T}))$ are in one-to-one relation (by the function Γ and ϕ whose restrictions are bijective).

<u>Proposition 6</u>.
The sets $\phi(\mathcal{P}(\mathbb{P}))$ and $\Gamma(\mathcal{P}(\mathbb{T}))$ are stable for the intersection . In fact, $\phi(\mathcal{P}(\mathbb{P}))$ is the closing of $\varphi(\mathbb{P})$ for the intersection and $\Gamma(\mathcal{P}(\mathbb{T}))$ is the one of $\gamma(\mathbb{T})$.

By an other way, the function Δ can be expressed using the function Γ :

Proposition 7.

$$\Delta(T) = \Gamma(T) - \cup_{T' \mid T \subset T' \subset \mathbb{T}, \ T' \neq T} \Gamma(T') = \Gamma(T) \cup_{t \in \mathbb{T} - T} \Gamma(T \cup \{t\}).$$

Representation of the elements of $\Gamma(\ \mathcal{P}(\mathbb{T}))$ by conjunction of linear constraints.
It is shown in [Fra,Sch,Taj:96] that :

Proposition 8.
$\forall\ t \in \mathbb{T}, \ \exists\ S \in \mathbb{C} \mid \gamma(t) = sol\ (S)$.

The sets P of form $\Gamma(T)$ verify $\Gamma(T) = \cap_{t \in T}\ \gamma(t)$ and so can also be specified by a conjunction of linear constraints :

Proposition 9.
$\forall\ T \in \mathcal{P}(\mathbb{T}), \ \exists\ S \in \mathbb{C} \mid \gamma(t) = sol\ (S)$.
The set of planes containing a given combination of tricubes may be specified by a conjunction of linear constraints.

Calculation of the set $\mathcal{T}$ of the effective tricubes.
Let $\mathcal{T}$ be the set of the values of the relation ρ.
$$\mathcal{T} = \{t \in \mathbb{T} \mid \exists p \in \mathbb{P}\ \ \rho(p,t)\} = \cup_{p \in \mathbb{P}} \mathcal{P}(p) \,.$$
The elements of $\mathcal{T}$, which are fourty, are the *effective tricubes* [Deb:95]. We can obtain these effective tricubes from a finite set of (potential) tricubes [6] by using Fourier's algorithm to the conjunction of linear constraints which correspond to them. (See [Fra,Sch,Taj:96]). Their list is given in the appendix A.

2.2 Calculation of the set of the complete combinations of multi-coplanar tricubes $\phi(\ \mathcal{P}(\mathbb{P}))$.

Calculation of the set $\Gamma(\ \mathcal{P}(\mathbb{T}))$ by means of an automaton.
The problem for the calculation of this set is the number of elements of $\mathcal{P}(\mathbb{T})$. Of course the calculation for the 2^{40} subsets of $\mathcal{T}$ is not viable ! Even the subsets containing at most nine tricubes are too numerous [7].
The adopted solution consists of calculating step by step, beginning with the empty set ($\Gamma(\varnothing) = \mathbb{P}$), and adding successively all the elements. So we build an automaton $\mathcal{A}$:

This (deterministic) automaton is specified by :
• its set of states : $\Gamma(\ \mathcal{P}(\mathbb{T}))$,　　($\subset \mathcal{P}(\mathbb{P})$)
• its alphabet : $\mathcal{T}$,

---

[6] It is easy to show that the effective tricubes verify some restrictions. For example, their image is included between -2 and +2. It is possible to refine more (see [Yac:97]).
[7] J.P.Reveillès [Rev:95] has shown that it was impossible to have more than nine different tricubes in a naive plane. (He shows a more general result, where some rectangles are put in place of the 3 × 3 squares of the tricubes).

- its function of transitions : $f:$ $\Gamma(\,\mathcal{P}(\mathbb{T})) \times \mathcal{T}$ $\to \Gamma(\,\mathcal{P}(\mathbb{T}))$,

$$\Gamma(T), t \qquad \dashrightarrow f(\Gamma(T), t) = \Gamma(T \cup \{t\}) = \Gamma(T) \cap \gamma(t),$$

- its initial state : $\mathbb{P}$,
- its set of final states : $\Gamma(\,\mathcal{P}(\mathbb{T})) - \varnothing$. [8]

The set of the states of the automaton is built from the initial state, by a repeated application of the transitions.

The states, represented by conjunctions of linear constraints, are stored in a list.

At the begining, the list contains only the initial state. Then, for each state figuring in the list, in order, one calculates the states to which leads the function of transition ; the so obtained states are introduced in the list, if they don't figure there yet. This process is performed until no new state appears.

This algorithm calculates $\Gamma(\,\mathcal{P}(\mathbb{T}))$; by a plaisant surprise, it ends quickly and gives the following result :

Proposition 10.
The set $\Gamma(\,\mathcal{P}(\mathbb{T}))$ has 148 elements (including $\mathbb{P}$ and $\varnothing$) .

Calculation of the set $\phi(\,\mathcal{P}(\mathbb{P}))$.
Thanks to the previously described bijection (proposition 5), the automaton $\mathcal{A}$ is able to be translated into an automaton $\mathcal{A}'$, whose states are sets of tricubes instead of sets of planes. In the usual practice, the two automatons are built simultaneously. The bijection[9] $\phi : \Gamma(\,\mathcal{P}(\mathbb{T})) \to \phi(\,\mathcal{P}(\mathbb{P}))$ is easily obtained from the automaton $\mathcal{A}$:
$\phi(\Gamma(T)) = \{t \in \mathcal{T} \mid f(\Gamma(T),t)=\Gamma(T)\}$ (set of tricubes for which the state is invariant) [10]

The previous result is immediately translated :

Theorem 1.
The set $\phi(\,\mathcal{P}(\mathbb{P}))$ of the complete combinations of multi-coplanar tricubes has 148 elements (including $\varnothing$ et $\mathbb{T}$) .

---

[8] The essential interest of this automaton is not the recognized language, but it is the construction of its set of states. It results that the final states have no importance.
[9] ϕ is not a bijection, but its restriction to the indicated sets is one. See above.
[10] Except for the empty state : $\phi(\varnothing) = \mathbb{T}$ and not $\mathcal{T}$, but it is without importance for the following.

2.3 Calculation of the set of complete combinations of coplanar tricubes $\varphi(\mathbb{P})$
$= \Delta^{-1}(\mathcal{P}(\mathbb{P})-\varnothing)$.

The set $\varphi(\mathbb{P})$ is obviously a subset of $\phi(\mathcal{P}(\mathbb{P}))$: one only has to restrict to the subsets of $\mathbb{P}$ with one element.

Let $T \in \phi(\mathcal{P}(\mathbb{P}))$: $T \in \varphi(\mathbb{P}) \Leftrightarrow \Delta(T) \neq \varnothing$.

We have shown that $\Delta(T) = \Gamma(T) - \cup_{i \in \mathbb{T} - T} \Gamma(T \cup \{t\})$ and we must test if this set is empty or not.

Therefore, we represent $\Delta(T)$ by a linear constraints system :

Let $S = \mathbb{T} - T$

$\Delta(T) = \Gamma(T) - \cup_{i \in S} (\Gamma(T) \cap \gamma(t)) = \Gamma(T) - \cup_{i \in S} \gamma(t)$

$\Delta(T) = \{p \mid p \in \Gamma(T) \wedge \neg (p \in \cup_{i \in S} \gamma(t)) \} = \{p \mid p \in \Gamma(T) \wedge (\wedge_{i \in S} \neg p \in \gamma(t)) \}$.

Let (c_i), (d_j), (e_j) be families of constraints such as

$p \in \Gamma(T) \Leftrightarrow \wedge_{i \in I} c_i$, $\qquad p \in \gamma(t) \Leftrightarrow \wedge_{j \in J(t)} d_j$, $\qquad e_j \Leftrightarrow \neg d_j$.

Then $\Delta(T) = \{p \mid (\wedge_{i \in I} c_i) \wedge (\wedge_{i \in S} \neg \wedge_{j \in J(t)} d_j)\}$.

We transform this expression to let appear a disjunction of constraints conjunctions :

$\Delta(T) = \{p \mid \vee_{f \in \prod_{i \in S} J(t)} ((\wedge_{i \in I} c_i) \wedge (\wedge_{i \in S} e_{f_i}))\}$.

To know if $\Delta(T)$ is empty, it is sufficient to apply Fourier's algorithm to each of the conjunctions. From a practical view, the applied transformation (distributivity) lets appear an exponential number of conjunctions : so it is convenient to use a variant optimized for the tranformation. The algorithm provides the following result :

Theorem 2.
The set $\varphi(\mathbb{P})$ of the complete combinations of coplanar tricubes has 99 elements.
The list of these combinations is given in the appendix B.

3. Conclusion.

This work has given an unexpected result : there are only 99 complete combinations of coplanar tricubes (for the naive planes whose normal is in the "first" 1/48 of space). It shows once more that the Fourier's algorithm can be, even in a practical way, efficient.

Currently, the use of this result is studied in LSIIT in Strasbourg to realise a polyhedrization of discrete volumes by facetization of their sides, that is to say by decomposing them into parts of discrete planes.

To pursue this work we consider the following subjects :
- a refinement of these coplanar tricubes combinations, which could show which of the tricubes are able to be adjoining and could help to understand the structure of the discrete planes.
- the generalisation to all planes (suppression of the 1/48 of space restriction).

I whish to thank J.Françon and M.Tajine for their help and suggestions, and for the fruitful discussions we had together.

Bibliography.

To have a description and a proof of the Fourier's algorithm one can consult [Kuh:56], or the chapter 1 "Inequalities Systems" of [Sto,Wit:70]. See also the bibliography from [Fra,Sch,Taj:96] . It includes references concerning the bases of arithmetic discret geometry, Fourier's algorithm as well as some more general results of calculability.

References.

[Deb:95] I. Debled-Rennesson. Etude et reconnaissance des droites et plans discrets. *Thèse de doctorat, Université Louis Pasteur,* Strasbourg, 1995.

[Deb,Rev:94] I. Debled-Rennesson, J.P. Reveillès. A new approach to digital planes. *Vision Geometry* III, Boston, 1994.

[Fra,Sch,Taj:96] J. Françon, J.M. Schramm, M. Tajine. Recognizing Arithmetic Straight Lines and Planes. *6th Conference on Discrete Geometry for Computer Imagery,* Lyon, 1996. Proceedings : Lecture Notes in Computer Science n°1176, Springer, 1996.

[Kuh:56] H. W. Kuhn. Solvability and consistency for linear equations and inequalities. *The Americain Mathematical Monthly,* vol. 63, p. 217-232, 1956.

[Rev:91] J.P. Reveillès. Géométrie discrète, calcul en nombres entiers et algorithmique. *Thèse de doctorat d'état, Université Louis Pasteur,* Strasbourg, 1991.

[Rev:95] J.P. Reveillès. Combinatorial pieces in digital lines and planes. *Vision Geometry 4 , SPIE'95,* San Diego, 1995.

[Sch:97] J.M.Schramm. Tricubes coplanaires. *RR97-12. LSIIT, Université Louis Pasteur,* Strasbourg, 1997.

[Sto-Wit:70] J. Stoer, C. Witzgall. Convexity and Optimization in Finite Dimensions I. Die Grundlehren der mathematischen Wissenschaften in Einzeldarstellungen, Band 163, Springer, 1970.

[Yac:97] J.Yaacoub. Enveloppes convexes de réseaux et applications au traitement d'images. *Thèse de doctorat, Université Louis Pasteur,* Strasbourg, 1997.

Appendix A : The forty effective tricubes.

tricube 1
```
          a>0
-1 -1 -1 | -a+b>0
 0  0 -1 | -a-b+c>0
 0  0  0 |
```

tricube 2
```
           a>=0
-1 -1 -1 | -2a+b>0
 0  0  0 | -2a-b+c>0
 0  0  0 |
```

tricube 3
```
          -a+b>=0
 0 -1 -1 | 2a-b>0
 0  0 -1 | -2b+c>0
 0  0  0 |
```

tricube 4
```
          -a+b>0
 0 -1 -1 | -2a-b+c>0
 0  0  0 | a>0
 0  0  0 | -2b+c>0
```

tricube 5
```
          -a+b>=0
 0  0 -1 | a>0
 0  0  0 | -a-2b+c>0
 0  0  0 |
```

tricube 6
```
           a>=0
 0  0  0 | -a+b>=0
 0  0  0 | -2a-2b+c>0
 0  0  0 |
```

tricube 7
```
           a+b-c>0
-1 -1 -2 | -a+b>0
 0  0 -1 | -b+c>0
 1  0  0 |
```

tricube 8
```
          -2a+c>0
-1 -1 -1 | -a-2b+2c>0
 0  0 -1 | 2a+b-c>0
 1  0  0 | 2b-c>0
```

tricube 9
```
          -2a+b>0
-1 -1 -1 | -a-b+c>0
 0  0  0 | 2b-c>0
 1  0  0 | a>0
```

tricube 10
```
          -a+b>=0
 0 -1 -2 | 2a-c>0
 0  0 -1 | a-2b+c>0
 1  0  0 |
```

tricube 11
```
          -a+b>=0
 0 -1 -1 | -a-2b+2c>0
         | 2a-b>0
 0  0 -1 | a-2b+c>0
 1  0  0 | 2a+b-c>0
```

tricube 12
```
          -a+b>0
 0 -1 -1 | -a-b+c>0
 0  0  0 | a-2b+c>0
 1  0  0 | a+2b-c>0
```

tricube 13
```
          -a+b>=0
 0  0 -1 | -2b+c>0
 0  0  0 | 2a+2b-c>0
 1  0  0 |
```

tricube 14
```
          -a+b>=0
 0  0  0 | -a-2b+c>0
 0  0  0 | a>0
 1  0  0 |
```

tricube 15
```
          -b+c>=0
-1 -1 -2 | a+2b-2c>0
 0  0 -1 | -a+2b-c>0
 1  1  0 |
```

tricube 16
```
          -b+c>0
-1 -1 -1 | -2a+c>0
 0  0 -1 | a+b-c>0
 1  1  0 | -a+2b-c>0
```

tricube 17
```
          -2a+b>0
-1 -1 -1 | -b+c>0
 0  0  0 | -a+2b-c>0
 1  1  0 | a>0
```

tricube 18
```
           2a-c>0
 0 -1 -2 | -b+c>0
 0  0 -1 | a+2b-2c>0
 1  1  0 | -a+b>0
```

tricube 19
```
          -2a-b+2c>0
 0 -1 -1 | 2a-b>0
 0  0 -1 | a+b-c>0
 1  1  0 | -a+b>0
```

tricube 20
```
          -2a+c>0
 0 -1 -1 | 2a-2b+c>0
 0  0  0 | 2b-c>0
 1  1  0 |
```

tricube 21
```
          -a-b+c>0
 0  0 -1 | a-2b+c>0
 0  0  0 | a+2b-c>0
 1  1  0 | -a+b>0
```

tricube 22
```
          -2a-b+c>0
 0  0  0 | -2b+c>0
 0  0  0 | -a+b>0
 1  1  0 | a>0
```

tricube 23
```
          -a+b>=0
 0 -1 -2 | 2a+b-2c>0
 1  0 -1 | -b+c>0
 1  1  0 |
```

tricube 24
```
          -a+b>=0
 0 -1 -1 | -2a-2b+3c>0
 1  0 -1 | 2a-c>0
 1  1  0 |
```

tricube 25
```
          -a+b>0
 0 -1 -1 | -2a-b+2c>0
 1  0  0 | a+b-c>0
 1  1  0 | 2a-b>0
```

tricube 26
```
          -a+b>=0
 0  0 -1 | -a-2b+2c>0
         | a-2b+c>0
 1  0  0 | 2a+b-c>0
 1  1  0 | 2a-b>0
```

tricube 27
```
          -a+b>=0
 0  0  0 | -2b+c>0
 1  0  0 | 2a-b>0
 1  1  0 |
```

tricube 28
```
          -a+b>=0
 0 -1 -2 | -b+c>=0
 1  0 -1 | 2a+2b-3c>0
 2  1  0 |
```

tricube 29
```
          -a+b>=0
 0 -1 -1 | -b+c>0
 1  0 -1 | 2a+b-2c>0
 2  1  0 |
```

tricube 30
```
          -a+b>0
 0 -1 -1 | -b+c>0
 1  0  0 | a+2b-2c>0
 2  1  0 | 2a-c>0
```

tricube 31
```
          -a+b>=0
 0  0 -1 | a-2b+c>0
 1  0  0 | 2a-c>0
 2  1  0 |
```

tricube 32
```
           a>=0
-1 -1 -1 | -b+c>=0
 0  0  0 | -2a+2b-c>0
 1  1  1 |
```

tricube 33
```
           a>0
 0 -1 -1 | -b+c>0
 0  0  0 | -a+2b-c>0
 1  1  1 | -2a+b>0
```

tricube 34
```
           a>0
 0  0 -1 | -a-b+c>0
 0  0  0 | 2b-c>0
 1  1  1 | -2a+b>0
```

tricube 35
```
           a>=0
 0  0  0 | -2a-b+c>0
 0  0  0 | -2a+b>0
 1  1  1 |
```

tricube 36
```
           a+b-c>0
 0 -1 -1 | -b+c>0
 1  0  0 | -a+2b-c>0
 1  1  1 | -2a+c>0
```

tricube 37
```
          -a-2b+2c>0
 0  0 -1 | 2a+b-c>0
 1  0  0 | 2b-c>0
 1  1  1 | -2a+c>0
```

tricube 38
```
          -a-b+c>0
 0  0  0 | a>0
 1  0  0 | -a+b>0
 1  1  1 |
```

tricube 39
```
          -b+c>=0
 0 -1 -1 | a+2b-2c>0
 1  0  0 | -a+2b-c>0
 2  1  1 |
```

tricube 40
```
          -b+c>0
 0  0 -1 | a+b-c>0
 1  0  0 | -a+b>0
 2  1  1 |
```

Appendix B : The 99 combinations of tricubes such as it exists a naive plane which contains all the tricubes from such a combination and no other tricube.

They are the elements from $\varphi(\mathbb{P})$.

```
{ 6}
{28}
{32}
{ 2,35}
{11,26}
{15,39}
{ 2, 6,35}
{ 2,32,35}
{ 3,13,27}
{ 8,20,37}
{10,24,31}
{15,28,39}
{15,32,39}
{ 1,12,21,38}
{ 3, 5,14,27}
{ 7,19,25,40}
{ 8,17,33,37}
{10,23,29,31}
{ 1, 4,13,22,38}
{ 1, 9,20,34,38}
{ 3, 5, 6,14,27}
{ 3, 5,13,14,27}
{ 3,11,13,26,27}
{ 7,16,20,36,40}
{ 7,18,24,30,40}
{ 8,11,20,26,37}
{ 8,17,20,33,37}
{ 8,17,32,33,37}
{10,11,24,26,31}
{10,23,24,29,31}
{10,23,28,29,31}
{ 1, 2,12,21,35,38}
{ 1, 4, 5,14,22,38}
{ 1, 9,17,33,34,38}
{ 1,11,12,21,26,38}
{ 7,11,19,25,26,40}
{ 7,15,19,25,39,40}
{ 7,16,17,33,36,40}
{ 7,18,23,29,30,40}
{ 1, 2, 4,13,22,35,38}
{ 1, 2, 9,20,34,35,38}
{ 1, 3, 4,13,22,27,38}
{ 1, 3,12,13,21,27,38}
{ 1, 4, 5, 6,14,22,38}
{ 1, 4, 5,13,14,22,38}
{ 1, 4,12,13,21,22,38}
{ 1, 8, 9,20,34,37,38}
{ 1, 8,12,20,21,37,38}
{ 1, 9,12,20,21,34,38}
{ 1, 9,17,20,33,34,38}
```

```
{ 1, 9,17,32,33,34,38}
{ 7, 8,16,20,36,37,40}
{ 7, 8,19,20,25,37,40}
{ 7,10,18,24,30,31,40}
{ 7,10,19,24,25,31,40}
{ 7,15,16,20,36,39,40}
{ 7,15,18,24,30,39,40}
{ 7,16,17,20,33,36,40}
{ 7,16,17,32,33,36,40}
{ 7,16,19,20,25,36,40}
{ 7,18,19,24,25,30,40}
{ 7,18,23,24,29,30,40}
{ 7,18,23,28,29,30,40}
{ 1, 2, 4, 5,14,22,35,38}
{ 1, 2, 9,17,33,34,35,38}
{ 1, 3, 4, 5,14,22,27,38}
{ 1, 8, 9,17,33,34,37,38}
{ 7, 8,16,17,33,36,37,40}
{ 7,10,18,23,29,30,31,40}
{ 7,15,16,17,33,36,39,40}
{ 7,15,18,23,29,30,39,40}
{ 1, 2, 4, 5, 6,14,22,35,38}
{ 1, 2, 4, 5,13,14,22,35,38}
{ 1, 2, 4,12,13,21,22,35,38}
{ 1, 2, 9,12,20,21,34,35,38}
{ 1, 2, 9,17,20,33,34,35,38}
{ 1, 2, 9,17,32,33,34,35,38}
{ 1, 3, 4, 5, 6,14,22,27,38}
{ 1, 3, 4, 5,13,14,22,27,38}
{ 1, 3, 4,12,13,21,22,27,38}
{ 1, 3,11,12,13,21,26,27,38}
{ 1, 8, 9,12,20,21,34,37,38}
{ 1, 8, 9,17,20,33,34,37,38}
{ 1, 8, 9,17,32,33,34,37,38}
{ 1, 8,11,12,20,21,26,37,38}
{ 7, 8,11,19,20,25,26,37,40}
{ 7, 8,16,17,20,33,36,37,40}
{ 7, 8,16,17,32,33,36,37,40}
{ 7, 8,16,19,20,25,36,37,40}
{ 7,10,11,19,24,25,26,31,40}
{ 7,10,18,19,24,25,30,31,40}
{ 7,10,18,23,24,29,30,31,40}
{ 7,10,18,23,28,29,30,31,40}
{ 7,15,16,17,20,33,36,39,40}
{ 7,15,16,17,32,33,36,39,40}
{ 7,15,16,19,20,25,36,39,40}
{ 7,15,18,19,24,25,30,39,40}
{ 7,15,18,23,24,29,30,39,40}
{ 7,15,18,23,28,29,30,39,40}
```

Coexistence of Tricubes in Digital Naive Plane

J. Vittone & J.M. Chassery

Laboratoire TIMC-IMAG, Institut Albert Bonniot
Domaine de la merci, 38706 La Tronche cedex, France
e-mail: {Joelle.Vittone,Jean-Marc.Chassery}@imag.fr

Abstract. Tricubes are considered as elementary 3D neighbours used to generate digital planes. We present some properties of these tricubes and discuss about their characterization and coexistence in a digital naive plane.

Key words. digital naive plane, bicube, tricube.

1 Introduction

Digital naive plane recognition is an important subject of research. Different works on this topic had been developped. An incremental algorithm has been introduced by Debled and Reveillès [DR95][DRR94] and more recently, works have been performed by Schramm based on Fourier's algorithm [FST96]. Here we are interested by the configurations of neighbourhoods, called tricubes, appearing in a digital naive plane. Properties about tricubes are introduced and we propose an array grammar based on tricubes for characterization of naive plane.

2 Definitions

A **digital naive plane**, 18-connected digital plane, is the set of points of $\mathbb{Z}^3$ satisfying the double inequality $\mu \leq ax + by + cz < \mu + max(|a|, |b|, |c|)$, $a, b, c, \mu \in \mathbb{Z}$, where (a, b, c) is the normal vector of the plane and μ a translation parameter. Using symetries, we can reduce our study to naive planes such as $0 \leq a \leq b \leq c$. We will note these planes $P(a, b, c, \mu)$.

The **lower** (resp. **upper**) **leaning points** of a naive plane $P(a, b, c, \mu)$ are the points verifying the equation $ax + by + cz = \mu$ (resp. $ax + by + cz = \mu + c - 1$).

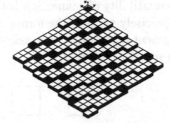

Fig. 1. *Naive plane $P(3, 7, 17, 0)$ on $[0, 14] \times [0, 14]$; in black upper leaning points, in gray lower leaning points.*

A **bicube** is a set of 4 voxels of a naive plane for which the projection on plane $(0xy)$ is reduced to a 2×2 square of pixels (fig.2).

Each voxel M in a naive plane P has 8 neighbours belonging to this plane. We call **tricube** attached to M in P the set of these 9 voxels. The projection of a tricube on the plane $(0xy)$ is reduced to a 3×3 square of pixels (fig.2).

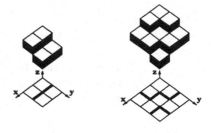

Fig. 2. *Example of bicubes and tricubes.*

3 Construction of tricubes by grouping bicubes

There exists 5 different configurations of bicubes (see fig.3) in a naive plane [Fra95][Fra96b][Fra96a].

Fig. 3. *The 5 different configurations of bicubes.*

Bicubes of a naive plane verify the following property:

Property 1. *[FST96] A naive plane can't contain simultaneously more than 4 different configurations of bicubes (types 0 and 4 can't appear in a same plane).*

A tricube can be seen as a grouping of four adjacent bicubes with recovering. We developed an algorithm to generate the set of tricubes. To do that we tested the compatibility of connection between bicubes.
More precisely we used the formal representation of bicubes of figure 4, where a full segment represents two voxels at same level and dot segment represents two voxels situated at different level.

Fig. 4. *Formal representation of the 5 bicubes.*

When we generate a tricube we note that two bicubes can connect only by a same type of line of voxels.

Now retaining only configurations verifying property 1, we obtain the 40 existing configurations of tricubes introduced by Debled[DR95] and illustrated in annexe.

4   Properties of a naive plane through its tricubes

We recall the following properties from previous works on digital planes:

Property 2. *[Rev95] There are no more than 9 different configurations of tricubes in a same naive plane.*

Property 3. *[Rev95] Each piece of plane which is projected as a 5 × 5 square on the plane (Oxy) and which is centered on a leaning point contains all the configurations of tricubes appearing in the plane.*

Let us introduce the following definitions:

Definition 1. *The* **symmetric of a tricube** *T is the tricube T' obtained by symetry around the central voxel of T. We notice by [T, T'] symetric tricubes. A tricube T is called* **neutral** *if T=T'. We notice by [T] the neutral tricubes. (see example on fig.5)*

(a)   (b)

Fig. 5. Example of: (a) symetric tricubes; (b)a neutral tricube.

and this lemma:

Lemma 1. *Let (x_l, y_l) be coordinates of the projection on plane (0xy) of a lower leaning point and (x_u, y_u) those of an upper leaning point of a same naive plane. Then for every $(\alpha, \beta) \in \mathbb{Z}^2$, the configuration of tricube at point $(x_l + \alpha, y_l + \beta)$ is symmetric from the configuration at point $(x_u - \alpha, y_u - \beta)$.*

we have finally:

Proposition 1. *If the number of configurations appearing in a naive plane is:*

- *odd then configurations of tricubes contain pairs of symmetric tricubes,*
- *even then one configuration is neutral and the others are pairs of symmetric tricubes.*

5 Grammar of tricubes

In order to enumerate digital planes we propose to use the number of tricubes involved in the decomposition of each plane. We have noticed that it exists two types of tricubes: the neutrals and the symetrics. In refering with annexe 1, the 6 neutral tricubes are numbered by [0], [8], [39], [19], [13] and [32]. We have seen that if a plane is generated with only one configuration of tricubes then this one corresponds to a neutral. Only [0], [8] and [39] can be used to generate a plane. These tricubes generate respectively the planes with normal vector $(0,0,1)$, $(0,1,1)$ and $(1,1,1)$. These planes define the limit of the part of space containing planes such as their normal vector (a,b,c) verify $0 \leq a \leq b \leq c$. So we will use them as a base on which we will construct other configurations.

Each normal vector (a,b,c) with $a,b,c \in \mathbb{Z}$ and $0 \leq a \leq b \leq c$ can be decomposed as $(c-b)(0,0,1)+(b-a)(0,1,1)+a(1,1,1)$. To generate all normal vectors of planes, we introduce a tree such as each node has three sons. At the top we start with vector (0,0,0). Then we generate the three sons of a node (x,y,z) in the following way:

- the left son is the vector $(x,y,z+1)$,
- the middle son is the vector $(x,y+1,z+1)$,
- the right son is the vector $(x+1,y+1,z+1)$.

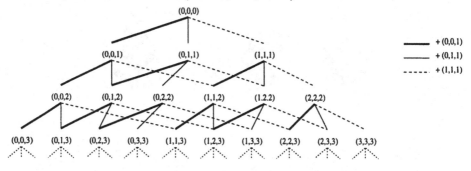

The development of grammar for 3D objects recognition is not recent. P.S.P. Wang [Wan89][SNW92][Wan92] had introduced a 3D array grammar to construct 3D objects using the notion of neighbourhood and adjancy. Here the study is limited to naive planes and we limit neighbourhood to tricubes. The grammar we propose is in the same time a geometric way to recognize naive planes but also an analytical model because it is linked to normal vector of naive planes.

Looking at the tricubes'table of Debled[DR95], we have for each tricube the normal vector (a,b,c) containing this tricube and such as c is minimal. We can remark that the 40 tricubes involve only 18 different normal vectors. Using the tree, we decompose these 18 vectors in a sum of 2 vectors (u_i, v_i, w_i) with $u_i \wedge v_i \wedge w_i = 1$. So we have the following table (at the right of the arrow we have the normal vector and its decomposition and at the left of the arrow, we have the list of tricubes which compose the plane):

$[0] \rightarrow (0,0,1)$
$[8] \rightarrow (0,1,1)$
$[39] \rightarrow (1,1,1)$
$[2,6] \rightarrow (0,1,2)=(0,0,1)+(0,1,1)$
$[24,30] \rightarrow (1,1,2)=(0,0,1)+(1,1,1)$
$[15,35] \rightarrow (1,2,2)=(0,1,1)+(1,1,1)$
$[5,29]+[19] \rightarrow (1,1,3)=(1,1,2)+(0,0,1)$
$[22,27]+[13] \rightarrow (1,2,3)=(0,1,2)+(1,1,1)=(1,2,2)+(0,0,1)=(1,1,2)+(0,1,1)$
$[25,36]+[32] \rightarrow (2,2,3)=(1,1,2)+(1,1,1)$
$[4,26]+[11,21] \rightarrow (1,2,4)=(1,2,3)+(0,0,1)=(0,1,2)+(1,1,2)=(1,1,3)+(0,1,1)$
$[23,34]+[16,31] \rightarrow (2,3,4)=(1,2,3)+(1,1,1)=(1,2,2)+(1,1,2)=(2,2,3)+(0,1,1)$
$[5,29]+[1,18] \rightarrow (1,1,4)=(1,1,3)+(0,0,1)$
$[22,27]+[9,12] \rightarrow (1,3,4)=(1,2,3)+(0,1,1)=(0,1,2)+(1,2,2)$
$[25,36]+[33,38] \rightarrow (3,3,4)=(2,2,3)+(1,1,1)$
$[3,10]+[19]+[4,26] \rightarrow (1,2,5)=(1,2,4)+(0,0,1)=(1,1,3)+(0,1,2)$
$[7,20]+[13]+[4,26] \rightarrow (1,3,5)=(1,2,4)+(0,1,1)=(1,2,3)+(0,1,2)$
$[14,28]+[13]+[23,34] \rightarrow (2,4,5)=(2,3,4)+(0,1,1)=(1,2,3)+(1,2,2)$
$[17,37]+[32]+[23,34] \rightarrow (3,4,5)=(2,3,4)+(1,1,1)=(2,2,3)+(1,2,2)$

From these formulae, we extract the following grammar:

$[0]+[0]=2[0]$
$[8]+[8]=2[8]$
$[39]+[39]=2[39]$
$[2,6]=[0]+[8]$
$[24,30]=[0]+[39]$
$[15,35]=[8]+[39]$
$[5,29]+[19]=[24,30]+[0]$
$[22,27]+[13]=[2,6]+[39]=[15,35]+[0]=[24,30]+[8]$
$[25,36]+[32]=[24,30]+[39]$
$[4,26]+[11,21]=[22,27]+[13]+[0]=[2,6]+[24,30]=[5,29]+[19]+[8]$
$[23,34]+[16,31]=[22,27]+[13]+[39]=[15,35]+[24,30]=[25,36]+[32]+[8]$
$[1,18]=[19]+[0]$
$[9,12]=[13]+[8]$ ou $[22,27]+[9,12]=[2,6]+[15,35]$
$[33,38]=[32]+[39]$
$[3,10]+[19]=[11,21]+[0]$ ou $[3,10]+[4,26]=[5,29]+[2,6]$
$[7,20]+[13]=[11,21]+[8]$ ou $[7,20]+[4,26]=[22,27]+[2,6]$
$[14,28]+[13]=[16,31]+[8]$ ou $[14,28]+[23,34]=[22,27]+[15,35]$
$[17,37]+[32]=[16,31]+[39]$ ou $[17,37]+[23,34]=[25,36]+[15,35]$

Let P be a naive plane with normal vector (a,b,c) with the conditions $0 \leq a \leq b \leq c$. Such a plane is represented by the point $(\frac{a}{c},\frac{b}{c})$ corresponding to the intersection of its normal vector with plane $(z=1)$. This point belongs to the triangle of vertices $A(0,0)$, $B(0,1)$ and $C(1,1)$ respectively associated to normal vectors $(0,0,1)$, $(0,1,1)$ and $(1,1,1)$.

Definition 2. *Let $P(a, b, c, \mu)$ be a naive plane identified by its tricubes $(T_1, T_2, ..., T_n)$, $n \leq 9$. We suppose $a \wedge b \wedge c = 1$ and $0 \leq a \leq b \leq c$.*
*P is an **exact plane** if it is the unique plane containing the tricubes $(T_1, T_2, ..., T_n)$.*

Example. The planes of normal vectors $(0, 0, 1)$, $(0, 1, 1)$ and $(1, 1, 1)$ contain respectively only one tricube [0], [8] and [39]. They are exact planes.

We propose the following algorithm to construct exact planes. It is based on **"barycentric construction"** of rationnal points.
Let P_1 be an exact plane identified by its normal vector (a_1, b_1, c_1) and its n_1 tricubes $(T_1, T_2, ..., T_{n_1})$.
Let P_2 be an exact plane identified by its normal vector (a_2, b_2, c_2) and its n_2 tricubes $(T'_1, T'_2, ..., T'_{n_2})$.
A new exact plane P_3 can be constructed from P_1 and P_2 if there exits a rule issued from the previous grammar which associates tricubes of P_1 with tricubes of P_2. In this case, the trace of the normal vector of the plane P_3 will be equal to the gravity center between trace of normal vectors of P_1 and P_2 weighted respectively by the number of tricubes appearing in the corresponding planes.

Example 1. Let P_1 be the plane of normal vector $(0, 0, 1)$ generated by the single tricube [0] and let P_2 be the plane of normal vector $(1, 1, 1)$ generated by the single tricube [39].
In figure 6(a), P_1 is illustrated by the point $(0, 0)$ and P_2 by the point $(1, 1)$.
Using rule [0]+[39]=[24,30], we can generate the exact plane P_3 such as the trace of its normal vector, here $(1, 1, 2)$, is equal to $\frac{1}{2}[(0, 0) + (1, 1)] = \left(\frac{1}{2}, \frac{1}{2}\right)$ and containing only tricubes [24,30] (see fig.6(b)).
The correspondance between $(1, 1, 2)$ and [24,30] is issued from the consultation of the first table (see also Annexe).

Example 2. Let P_1 be the plane of normal vector $(0, 0, 1)$ generated by the single tricube [0] and let P_2 be the plane of normal vector $(1, 1, 2)$ generated by tricubes [24,30].
In figure 6(b), P_1 is illustrated by the point $(0, 0)$ and P_2 by the point $\left(\frac{1}{2}, \frac{1}{2}\right)$.
Using rule [24,30]+[0]=[5,29]+[19], we can generate the exact plane P_3 such as the trace of its normal vector $(1, 1, 3)$ is equal to $\frac{1}{2+1}[(0, 0) + \left(\frac{1}{2}, \frac{1}{2}\right)] = \left(\frac{1}{3}, \frac{1}{3}\right)$ and containing tricubes [5,29][19] (see fig.6(c)).
The correspondance between $(1, 1, 3)$ and $5, 29, 19$ is issued from the consultation of table 1 (see Annexe).
We can note in figure 6(c) that tricubes [8] and [2,6] can't be associated to generate an exact plane because there is no rule between [8] and [2,6].

Using such algorithm we generate all the exact planes (see fig.6).

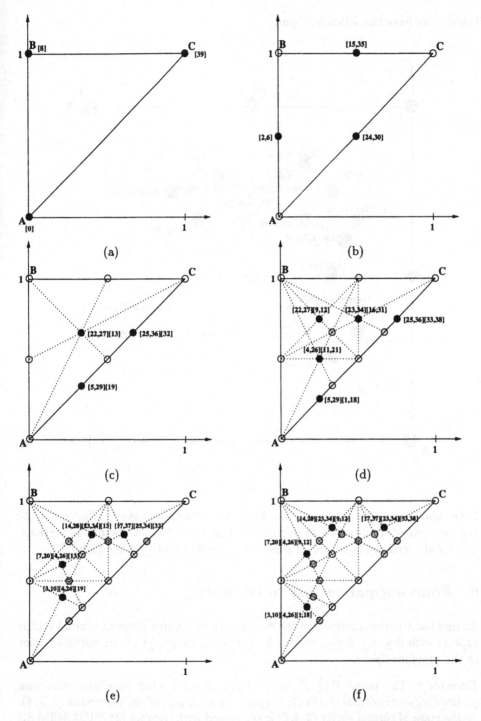

Fig. 6. Construction of the projection on plane ($z = 1$) of the normal vectors of the exact planes containing 1(a), 2(b), 3(c), 4(d), 5(e) and 6(f) tricubes.

Finally, we have the following figure:

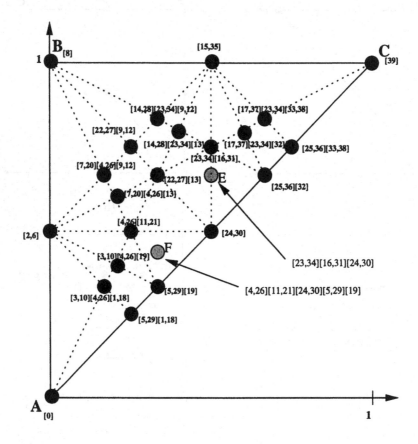

In the above figure, in addition to the 22 "exact" points we have 49 edges connecting two "exact" points and 28 polygonal areas with " exact" points as vertices. The total corresponds to the 99 possible associations of tricubes [Sch97].

6 From normal vectors to tricubes

To find the tricubes involved in the description of a naive plane of normal vector (a, b, c) with $0 \leq a \leq b \leq c$, we can locate the trace $\left(\frac{a}{c}, \frac{b}{c}\right)$ of the normal vector in the previous figure.

Example 1. The point $E\left(\frac{3}{6} \cdot \frac{4}{6}\right)$ in the figure is supported by edge connecting planes $P_{[24,30]}$ of normal vector $(1, 1, 2)$ and $P_{[23,34][16,31]}$ of normal vector $(2, 3, 4)$. So the plane of normal vector $(3, 4, 6)$ is composed with tricubes [24,30][23,34][16,31].

Example 2. The point $F\left(\frac{3}{9} \cdot \frac{4}{9}\right)$ in the figure is in the triangular area which vertices are planes $P_{[24,30]}$ of normal vector $(1, 1, 2)$, $P_{[4,26][11,21]}$ of normal vector

$(1, 2, 4)$ and $P_{[5,29][19]}$ of normal vector $(1, 1, 3)$. So the plane of normal vector $(3, 4, 9)$ is composed with tricubes $[24,30][4,26][11,21][5,29][19]$.

Here we propose an algorithm which takes into account the position of the trace into the triangle ABC. This position will be compared with positions of tricubes $[0]$, $[8]$, $[39]$, $[24,30]$, $[2,6]$ and $[15,35]$.

```
- Initialization:
        x = c − b
        y = b − a
        z = a
- Decompose in x[0] + y[8] + z[39]
- Make appear [24,30]:
```
$\quad\quad k_1[24, 30] + (x - k_1)[0] + y[8] + (z - k_1)[39]$ with $k_1 = min(x, z)$

If $k_1 = z$ **Then** {no more [39]}

$\quad$ - Make appear [2,6]:

$\quad\quad k_2[2, 6] + k_1[24, 30] + (x - k_1 - k_2)[0] + (y - k_2)[8]$ with $k_2 = min(y, x - k_1)$

$\quad$ **If** $k_2 = x - k_1$ **Then** {no more [0]}

$\quad$ - We have the solution in calculating:

$k_6[9, 12] + k_5[7, 20] + k_4[22, 27] + (k_4 + k_5 - k_6)[13] + k_3[4, 26] + (k_3 - k_5)[11, 21]$
$\quad + (k_2 - k_3)[2, 6] + (k_1 - k_3 - k_4)[24, 30] + (y - k_2 - k_4 - k_5 - k_6)[8]$
$\quad\quad$ with $val = y$

$\quad$ **Else** {no more [8]}

$\quad$ - We have the solution in calculating:

$k_6[1, 18] + k_5[3, 10] + k_4[5, 29] + (k_4 + k_5 - k_6)[19] + k_3[4, 26] + (k_3 - k_5)[11, 21]$
$\quad + (k_2 - k_3)[2, 6] + (k_1 - k_3 - k_4)[24, 30] + (x - k_1 - k_2 - k_4 - k_5 - k_6)[0]$
$\quad\quad$ with $val = x - k_1$

Else {no more [0]}

$\quad$ - Make appear [15,35]:

$\quad\quad k_2[15, 35] + k_1[24, 30] + (z - k_1 - k_2)[39] + (y - k_2)[8]$ with $k_2 = min(y, z - k_1)$

$\quad$ **If** $k_2 = z - k_1$ **Then** {no more [39]}

$\quad$ - We have the solution in calculating:

$k_6[9, 12] + k_5[14, 28] + k_4[22, 27] + (k_4 + k_5 - k_6)[13] + k_3[23, 34] + (k_3 - k_5)[16, 31]$
$\quad + (k_2 - k_3)[15, 35] + (k_1 - k_3 - k_4)[24, 30] + (y - k_2 - k_4 - k_5 - k_6)[8]$
$\quad\quad$ with $val = y$

$\quad$ **Else** {no more [8]}

$\quad$ - We have the solution in calculating:

$k_6[33, 38] + k_5[17, 37] + k_4[25, 36] + (k_4 + k_5 - k_6)[32] + k_3[23, 34] + (k_3 - k_5)[16, 31]$
$\quad + (k_2 - k_3)[15, 35] + (k_1 - k_3 - k_4)[24, 30] + (z - k_1 - k_2 - k_4 - k_5 - k_6)[39]$
$\quad\quad$ with $val = z - k_1$

with

$$k_3 = min(k_1, k_2)$$
$$k_4 = min(k_1 - k_3, val - k_2)$$
$$k_5 = min(k_3, val - k_2 - k_4)$$
$$k_6 = min(k_4 + k_5, val - k_2 - k_4 - k_5)$$

Example. If we examine the normal vector $(3,5,8)$. This vector can be decomposed as $3(0, 0, 1) + 2(0, 1, 1) + 3(1, 1, 1)$. So we have $3[0] + 2[8] + 3[39]$. With the algorithm, we obtain $[24, 30] + 2[22, 27] + 2[13]$. The plane P(3,5,8,0) is formed with tricubes 22, 27, 13, 24 and 30. Figure 7 illustrates the configurations of tricubes around a leaning point of this same plane (see property 3).

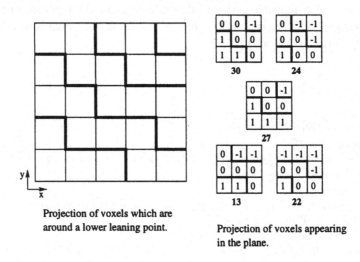

Projection of voxels which are around a lower leaning point.

Projection of voxels appearing in the plane.

Fig. 7. *Configurations of tricubes appearing in plane* $P(3, 5, 8, 0)$.

7 Conclusion

This work describes the way to associate normal vector of a plane and basic elements (tricubes). From a list of tricubes appearing in a set of voxels near leaning points, we are able to say if these tricubes can coexist in a naive plane. A geometric model to recognize naive digital plane has been presented and we retrieve results issued from other analytic method [Sch97].

In perspective we are working on the polyhedrization of a set of voxels and on relations between our coding of tricubes and the umbrella graph presented by Françon [Fra95].

References

[BF94]   Ph. Borianne and J. Françon. Reversible polyhedrization of discrete volumes. In *Proc. DCGI'4*, pages 157–168, Grenoble,France, September 1994.

[CM91]   J.M. Chassery and A. Montanvert. *Géométrie discrète en analyse d'images*, page 456. Hermès, Paris, 1991.

[DR95]   I. Debled-Renesson. *Etude et reconnaissance des droites et plans discrets.* PhD thesis, University Louis Pasteur, Strasbourg,France, 1995.

[DRR94] I. Debled-Renesson and J.P. Reveillès. An incremental algorithm for digital plane recognition. In *Proc. DCGI'4*, pages 207–222, Grenoble,France, September 1994.

[Fra95] J. Françon. Arithmetic planes and combinatorial manifolds. In *Proc. DCGI'5*, pages 209–217, Clermont-Ferrand,France, September 1995.

109

[Fra96a] J. Françon. On recent trends in discrete geometry in computer science. In *Lecture Notes in Computer Science*, volume 1176, pages 3–16. S. Miguet, A. Montanvert and S. Ubéda Eds, Springer, 1996.

[Fra96b] J. Françon. Sur la topologie d'un plan arithmétique. In *Theorical Computer Science*, volume 156, pages 159–176. Elsevier, 1996.

[FST96] J. Françon, J.M. Schramm, and M. Tajine. Recognizing arithmetic straight lines and planes. In *Lecture Notes in Computer Science*, volume 1176, pages 141–150. S. Miguet, A. Montanvert and S. Ubéda Eds, Springer, 1996.

[Rev95] J.P. Reveillès. Combinatorial pieces in digital lines and planes. In *Vision Geometry IV*, volume 2573. SPIE, 1995.

[Sch97] J.M. Schramm. Tricubes coplanaires. Technical report, University of Strasbourg, France, Fev. 1997.

[SNW92] A. Saoudi, M. Nivat, and P.S.P. Wang(Eds.). Parallel image processing. In *International journal of pattern recognition and artificial intelligence*, volume 6. WSP, 1992.

[Wan89] P.S.P. Wang(Ed.). Array grammars, patterns and recognizers. In *International journal of pattern recognition and artificial intelligence*, volume 3. WSP, 1989.

[Wan92] P.S.P. Wang. Parallel image analysis : proceedings. In *ICPIA'92, Lecture Notes in Computer Science*, volume 654. A. Nakamura, M. Nivat, A. Saoudi, P.S.P. Wang and K. Inoue (Eds.), Springer, 1992.

Annexe: The 40 configurations of tricubes

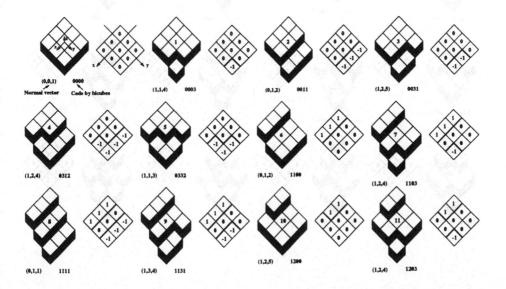

(1,3,4) 1211   (1,2,3) 1231   (2,4,5) 1412   (1,2,2) 1414

(2,3,4) 1432   (3,4,5) 1434   (1,1,4) 2000   (1,1,3) 2003

(1,3,5) 2011   (1,2,4) 2031   (1,2,3) 2312   (2,3,4) 2314

(1,1,2) 2332   (2,2,3) 2334   (1,2,4) 3120   (1,2,3) 3123

(2,4,5) 3141   (1,1,3) 3220   (1,1,2) 3223   (2,3,4) 3241

(2,2,3) 3442   (3,3,4) 3444   (2,3,4) 4123   (1,2,2) 4141

(3,3,4) 4223   (3,4,5) 4241   (3,3,4) 4442   (1,1,1) 4444

Surfaces

Some Structural Properties of Discrete Surfaces

Gilles Bertrand and Michel Couprie

Laboratoire PSI, ESIEE Cité Descartes B.P. 99
93162 Noisy-Le-Grand Cedex France, e-mail: bertrang@esiee.fr, coupriem@esiee.fr

Abstract. In the framework of combinatorial topology a surface is described as a set of faces which are linked by adjacency relations. This corresponds to a structural description of surfaces where we have some desirable properties: for example, any point is surrounded by a set of faces which constitute a "cycle". The notion of combinatorial surface extracts these "structural" properties of surfaces.

In this paper, we introduce a relation for points in Z^3 which is based on the notion of homotopy. This allows to propose a definition of a class of surfaces which are combinatorial surfaces. We then show that the main existing notions of discrete surfaces belong to this class of combinatorial surfaces.

Keywords: surfaces, discrete topology, homotopy, simple points

1 Introduction

In the three-dimensional discrete space Z^3, several approaches of surfaces have been proposed:

- a graph-theoretical approach: a surface is defined as a set of points linked by adjacency relations [16, 17, 20];
- a voxel approach: a surface is defined as a set of faces (surfels) between pairs of adjacent voxels [1, 8];
- a general topology approach [13];
- a combinatorial approach: a surface is defined as a structure [7, 9, 15].

In the framework of combinatorial topology a surface is described as a set of faces which are linked by adjacency relations. This corresponds to a structural description of surfaces where we have some desirable properties: for example, any point is surrounded by a set of faces which constitute a "cycle". The notion of combinatorial surface extracts these "structural" properties of surfaces.

The graph-theoretical definitions of closed surfaces are not based upon structural properties. In fact, the structural nature of these surfaces is difficult to extract. The major problem which arises for these surfaces is that the adjacency relation used for defining them does not induce a structural relation. For example, the neighborhood of a point does not constitute a simple closed curve under the adjacency relation.

In this paper, we make a link between the definitions of surfaces based on the graph-theoretical approach and the combinatorial approach. For that purpose, we introduce a relation for points in Z^3 which is based on the notion of homotopy. This allows to propose a definition of a class of surfaces which are combinatorial surfaces. We then show that the main existing notions of surfaces belong to this class of combinatorial surfaces.

2 Basic notions

We recall some basic notions of 3D discrete topology (see also [12]).

We denote $E = Z^3$, Z being the set of relative integers. A point $x \in E$ is defined by (x_1, x_2, x_3) with $x_i \in Z$. We consider the four neighborhoods:

$N_{124}(x) = \{x' \in E; \; Max[|x_1 - x'_1|, |x_2 - x'_2|, |x_3 - x'_3|] \leq 2\}$,

$N_{26}(x) = \{x' \in E; \; Max[|x_1 - x'_1|, |x_2 - x'_2|, |x_3 - x'_3|] \leq 1\}$,

$N_{18}(x) = \{x' \in E; \; |x_1 - x'_1| + |x_2 - x'_2| + |x_3 - x'_3| \leq 2\} \cap N_{26}(x)$,

$N_6(x) = \{x' \in E; \; |x_1 - x'_1| + |x_2 - x'_2| + |x_3 - x'_3| \leq 1\}$.

We define $N_k^*(x) = N_k(x) \setminus \{x\}$, with $k = 6, 18, 26, 124$.

Two points x and y are said to be n-adjacent (n = 6, 18, 26) if $y \in N_n^*(x)$; we also say that y is an n-neighbor of x.

We denote $N_{18}^+(x) = N_{18}^*(x) \setminus N_6^*(x)$ and $N_{26}^+(x) = N_{26}^*(x) \setminus N_{18}^*(x)$.

Two points x and y are said to be strictly n-adjacent (n = 18, 26) if $y \in N_n^+(x)$.

An n-path π is a (possibly empty) sequence of points $x_0..x_k$, with x_i n-adjacent to x_{i-1}, for $i = 1..k$. If π is not empty, the length of π is equal to k. If $x_0 = x_k$, π is closed and x_0 is called the origin of π.

Let X be a subset of E. We denote by $\overline{X}$ the complement of X.

Let $x \in X$ and $y \in X$. We say that x is n-connected to y if there is an n-path in X between x and y. The relation "is n-connected to" is an equivalence relation. The equivalence classes relative to this relation are the n-connected components of X (or simply the n-components of X).

A subset X of E is n-connected if it is made of exactly one n-connected component.

A subset X of E is a simple closed n-curve if X is n-connected and if each point of X is n-adjacent to exactly two points in X.

As in 2D, if we use an n-adjacency relation for X we have to use another $\overline{n}$-adjacency relation for $\overline{X}$, i.e. the 6-adjacency for X is associated to the 18- or the 26-adjacency for $\overline{X}$ (and vice versa). This is necessary for having a correspondence between the topology of X and the topology of $\overline{X}$. Furthermore, it is sometimes necessary to distinguish the 6-adjacency associated with the 18-adjacency and the 6-adjacency associated with the 26-adjacency. Whenever we will have to make this distinction, a 6^+-notion will indicate a 6-notion associated with the 18-adjacency. So, we can have $(n, \overline{n}) = (6, 26), (26, 6), (6^+, 18)$ or $(18, 6^+)$.

Note that, if X is finite, the infinite $\overline{n}$-connected component of $\overline{X}$ is the background, the other $\overline{n}$-connected components of $\overline{X}$ are the cavities.

The notion of deformation allows to detect the presence of a "hole" in a set X (see [10]).

Let $X \subset E$ and let $p \in X$ be a point, called the base point. Let γ and γ' be two closed n-paths composed of points of X and which have p as origin. We say that γ' is *an elementary n-deformation* of γ, or $\gamma \sim \gamma'$, if there are two n-paths π_1, π_2, and two non-empty n-paths π, π', such that γ and γ' are of the form $\gamma = \pi_1\pi\pi_2$, $\gamma' = \pi_1\pi'\pi_2$, and such that all points of π and π' are included in a little portion P of E:

- for $n = 6$, P is a unit square (a 2×2 square);
- for $n = 6^+$, 18, 26, P is a unit cube (a $2 \times 2 \times 2$ cube).

We say that γ' is an *n-deformation* of γ or $\gamma \simeq \gamma'$ if there is a sequence of closed n-paths $\gamma_0..\gamma_k$ such that $\gamma = \gamma_0$, $\gamma' = \gamma_k$ and $\gamma_{i-1} \sim \gamma_i$ for $i = 1..k$.

Let $\gamma = px_0...x_ip$ and $\gamma' = px'_0...x'_jp$ be two closed n-paths composed of points of X and which have p as origin. The *product* of γ and γ' is the closed n-path $px_0...x_ipx'_0...x'_jp$ obtained by catenating γ and γ'.

Let us consider the classes of equivalence of the closed n-paths with origin p under the relation $\simeq$. We may define the product of two such classes as the equivalence class of the product of two closed n-paths corresponding to the classes.

Under the product operation, these classes constitute a group $\Pi_n(p, X)$ which is the *fundamental n-group* (or Poincaré group) with base point p. As in the continuous spaces, the fundamental group reflects the structure of the holes (or tunnels) in X. For example, the fundamental group of a hollow torus is a free abelian group on two generators. Note that if p and q belong to the same n-connected component of X, then $\Pi_n(p, X)$ is isomorphic to $\Pi_n(q, X)$.

3 Homotopy and strong homotopy

In this section, we recall some notions of homotopy and strong homotopy. The homotopy in a discrete grid may be defined through the notion of simple point (see also [10]).

Let $n \in \{6, 6^+, 18, 26\}$. Let $X \subset E$. A point $x \in E$ is said to be *n-simple (for X)* if its removal from X (if $x \in X$) or its addition to X (if $x \in \overline{X}$) does not "change the topology of the image", i.e., if:

    1) There is a one to one correspondence between the n-connected components of $X \setminus \{x\}$ and the n-connected components of $X \cup \{x\}$; and

    2) There is a one to one correspondence between the $\overline{n}$-connected components of $\overline{X} \setminus \{x\}$ and the $\overline{n}$-connected components of $\overline{X} \cup \{x\}$; and

    3) For each point p of $X \setminus \{x\}$, the inclusion map $i: X \setminus \{x\} \to X \cup \{x\}$ induces a group isomorphism $i^*: \Pi_n(p, X \setminus \{x\}) \to \Pi_n(p, X \cup \{x\})$; and

    4) For each point q of $\overline{X} \setminus \{x\}$, the inclusion map $j: \overline{X} \setminus \{x\} \to \overline{X} \cup \{x\}$ induces a group isomorphism $j^*: \Pi_{\overline{n}}(q, \overline{X} \setminus \{x\}) \to \Pi_{\overline{n}}(q, \overline{X} \cup \{x\})$.

The set $Y \subset X$ is *lower n-homotopic* to X if there exists a sequence of sets $Z_0, ..., Z_k$, with $Z_0 = Y$, $Z_k = X$, such that $Z_{i-1} \subset Z_i$ and $Z_i \setminus Z_{i-1}$ consists in a single point which is an n-simple point for Z_{i-1}, $i = 1, ..., k$. The set $S \subset X$ is called a *(lower) n-simple set for X*, if $X \setminus \{S\}$ is lower n-homotopic to X.

Thus, the set Y is lower n-homotopic to X, if X may be obtained from Y by iterative additions of n-simple points, or, equivalently, if Y may be obtained from X by iterative deletions of n-simple points.

Let $X \subset E$ and $x \in E$. We denote $|X|^x = N_{26}^*(x) \cap X$.

The *geodesic n-neighborhood of x inside X of order k* is the set $N_n^k(x, X)$ defined recursively by: $N_n^1(x, X) = N_n^*(x) \cap X$ and $N_n^k(x, X) = \cup\{N_n(y) \cap |X|^x, y \in N_n^{k-1}(x, X)\}$.

In other words $N_n^k(x, X)$ is the set composed of all points y of $|X|^x$ such that there exists an n-path π from x to y of length less than or equal to k, all points of π, except x, belonging to $|X|^x$. We give now a definition of topological numbers which leads to a characterization of simple points [2]:

Definition 1: Let $X \subset E$, $x \in E$ and $n \in \{6, 6^+, 18, 26\}$.

The *geodesic neighborhoods* $G_n(x, X)$ are defined by:

$$G_6(x, X) = N_6^2(x, X); \quad G_{6^+}(x, X) = N_6^3(x, X);$$
$$G_{18}(x, X) = N_{18}^2(x, X); \quad G_{26}(x, X) = N_{26}^1(x, X).$$

The *topological number* $T_n(x, X)$ is defined as the number of n-components in $G_n(x, X)$.

Note that the topological number depends only on the neighborhood $N_{26}^*(x) \cap X$, we have $T_n(x, X) = T_n(x, |X|^x)$. The evaluation of the topological number may be done by using classical graph-theoretic algorithms for searching connected components. We have [2]:

Theorem 2: Let $X \subset E$ and $x \in E$: x is an n-simple point if and only if $T_n(x, X) = 1$ and $T_{\overline{n}}(x, \overline{X}) = 1$.

We introduce the notion of strong homotopy: see [3], see also the work of Kong [11] in which the notion of hereditarily simple set is introduced, this notion is equivalent to the notion of strongly simple set presented hereafter:

Definition 3: Let $X \subset E$ and $Y \subset X$. The set Y is *strongly (lower) n-homotopic to* X if, for each subset Z such that $Y \subset Z \subset X$, Z is lower n-homotopic to X.

If Y is strongly n-homotopic to X, we say that $X \setminus Y$ is a *strongly (lower) n-simple set.*

4 Existing notions of surfaces

We now present existing notions of surfaces. First of all, we give some general definitions.

Definition 4: Let X be a subset of E and let x be a point of E.

The set X is an *n-thin set* if, $\forall x \in X$, $|\overline{X}|^x$ has exactly two $\overline{n}$-components which are $\overline{n}$-adjacent to x.

The set X is an *n-Jordan set* if X is n-connected and if $\overline{X}$ has two $\overline{n}$-connected components.

If X is an n-Jordan set, we will denote by A and B the two components of $\overline{X}$. These components are called the *back-components* of X. The *closure* of a back-component is the union of this back-component and X.

An *n-separating set* is an n-Jordan set which is also an n-thin set.

An n-separating set X is a *strongly n-separating set* if, $\forall x \in X$, x is $\overline{n}$-adjacent to both A and B.

4.1 Morgenthaler's surfaces

Let us present the definition of surfaces introduced by Morgenthaler and Rosenfeld [20]:

Definition 5: Let $(n, \overline{n}) = (6, 26)$ or $(26, 6)$ and let X be a subset of E. A point x of X is a *Morgenthaler's (simple) n-surface point* if:

1) $|X|^x$ has exactly one n-component which is n-adjacent to x; and

2) $|\overline{X}|^x$ has exactly two $\overline{n}$-components which are $\overline{n}$-adjacent to x; we denote C^{xx} and D^{xx} these components; and

3) $\forall y \in N_n(x) \cap X$, $N_{\overline{n}}(y) \cap C^{xx} \neq \emptyset$ and $N_{\overline{n}}(y) \cap D^{xx} \neq \emptyset$.

Furthermore, if $N_{124}(x) \cap \overline{X}$ has exactly two $\overline{n}$-components which are $\overline{n}$-adjacent to x, we say that the n-surface point x is *orientable*.

A *Morgenthaler's (simple) closed n-surface* is a finite n-connected set X consisting entirely of orientable Morgenthaler's n-surface points.

In Fig. 1, the configuration (a) corresponds to a Morgenthaler's 6-surface point, the configurations (b), (e) correspond to Morgenthaler's 26-surface points. It was shown that [20]:

Theorem 6: *A Morgenthaler's closed n-surface is a strongly n-separating set.*

Furthermore, it was proved that the assumption of orientability is unnecessary for the 6-connectivity [22] and for the 26-connectivity [21].

4.2 Malgouyres' surfaces

Malgouyres' surfaces [17] are based on a generalization of the notion of a simple closed curve.

Definition 7: Let X be a subset of E. We say that a point x of X is an *n-corner* if x is n-adjacent to two and only two points y and z belonging to X such that y and z are themself n-adjacent; we say that the n-corner x is *simple* if y and z are not corners and if x is the only point n-adjacent to both y and z. We say that X is a *generalized simple closed n-curve*, or a *G_n-curve*, if the set obtained by removing all simple n-corners of X is a simple closed n-curve.

Definition 8: A finite subset X of E is called a *Malgouyres' (simple) closed 18-surface* if X is 18-connected and if, for each x of X, the set $|X|^x$ is a G_{18}-curve.

In Figure 1, (b), (c), (f), (g) are examples of Malgouyres' surface points. It was proved in [17] that:

Theorem 9: *Any Malgouyres' 18-surface is a strongly n-separating set, for $n = 18$ and $n = 26$.*

4.3 Strong surfaces

The definition of strong surfaces is based on the notion of strong homotopy (see [5]):

Definition 10: Let $X \subset E$ be an n-separating set. The set X is a *strong (closed) n-surface* if any back-component of X is strongly n-homotopic to its closure.

We have the following result [5]:

Theorem 11: *Let $X \subset E$ be an n-separating set.*
X is a strong n-surface if and only if, for each x of X, each of the four following conditions is satisfied:

1) $T_n(x, |A|^x) = 1$ and $T_n(x, |B|^x) = 1$;
2) $T_{\overline{n}}(x, |A|^x) = 1$ and $T_{\overline{n}}(x, |B|^x) = 1$;
3) $\forall y \in N_n^*(x) \cap X$, $T_n(x, |A|^x \cup \{y\}) = 1$ and $T_n(x, |B|^x \cup \{y\}) = 1$;
4) $\forall y \in N_{\overline{n}}^*(x) \cap X$, $T_{\overline{n}}(x, |A|^x \cup \{y\}) = 1$ and $T_{\overline{n}}(x, |B|^x \cup \{y\}) = 1$.

In Fig. 1, the central points of (b), (c), (d), (g), could satisfy the conditions of Th. 11 for strong 26-surfaces. The central points of (b), (c), (d), (e), (f), (g), (h), could satisfy the conditions of Th. 11 for strong 18-surfaces.

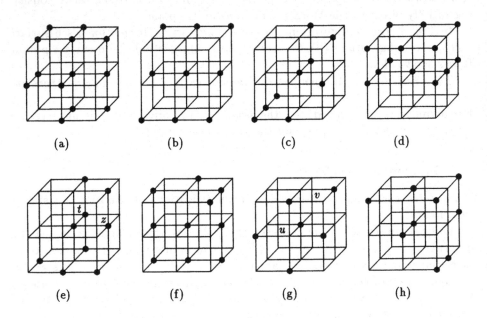

Fig. 1. Examples.

A fully local characterization of strong 26-surfaces was proposed [6], [19]. Furthermore, we have [5]:

Theorem 12: *Any Morgenthaler's closed 26-surface is a strong 26-surface.*
Theorem 13: *Any Malgouyres' closed 18-surface is a strong 18-surface.*

5   Combinatorial manifold

We introduce the notion of two-dimensional combinatorial manifold. We use the same definitions as in [7].

Definition 14: Let G be a graph. An *oriented loop* of G is a circular permutation $L = (v_0, v_1, .., v_{k-1})$, $k > 2$, of vertices of G such that , for all i, v_i is adjacent to v_{i+1} (indices taken modulo k) and $v_i \neq v_j$ if $i \neq j$; a *loop* of G is an oriented loop up to its orientation. A vertex v_i of a loop is called *adjacent* to the loop; the oriented edge (v_i, v_{i+1}) (resp. the edge $\{v_i, v_{i+1}\}$) is called *adjacent* to the oriented loop (resp. the loop). Two loops having a common edge are called adjacent.

Definition 15: A *two dimensional (closed) combinatorial manifold* $M = [G, F]$ is a graph G together with a set F of loops of G, called *faces* or *2-cells* of M, such that:
1) every edge of G is adjacent to exactly two faces, and
2) for every vertex v, the set of faces adjacent to v can be organized in a circular permutation $(f_0, f_1, ..., f_{k-1})$, $k > 1$, called the *umbrella* of v, such that, for all i, f_i is adjacent to f_{i+1} (indices taken modulo k).
The vertices (resp. edges) of a combinatorial manifold are also called the *0-cells* (resp. *1-cells*).

The notion of 2D combinatorial manifold corresponds to a structural description of a surface. It is then desirable to have such a description for surfaces in Z^3. The major problem is that the n-adjacency relation used to define these surfaces does not allow to recover this description. Let us see for example the simple configuration depicted Fig. 2 (a). It corresponds to a Morgenthaler's 26-surface point. We see that, under the 26-adjacency relation, the set of points of the surface which surround the central point x does not constitute a simple closed curve. For example the point 1 has four 26-neighbors $(7, 8, 2, 3)$ in $N_{26}^*(x)$. The 26-adjacency relations for the configuration of Fig. 2 (a) are depicted Fig. 2 (b). We see that the elementary loops for the graph corresponding to the 26-adjacency does not satisfy the conditions of Def. 15: for example, the edge $\{x, 3\}$ is adjacent to the four loops $(x, 2, 3)$, $(x, 1, 3)$, $(x, 3, 4)$, $(x, 3, 5)$.

Thus, we have to consider another relation for extracting the structure of such surfaces. We consider a relation based upon the notion of homotopy and simple point. Let us consider again the configuration of Fig. 2 (a). We suppose that all the neighbors of the central point x have a 26-neighborhood which also corresponds to this configuration, i.e., the $5 \times 5 \times 5$ neighborhood of x is a digital plane. Let us first note that all points of a surface are non-simple points. Suppose now that the point x is removed; we see that there are four points 1, 3, 5, 7 which will appear as 26-simple points; the other points will not appear as simple, for example the neighborhood of the point 2 after deletion of x is depicted Fig. 2 (c), this does not correspond to a 26-simple point. Let us denote $S(x)$ the set of points which appear as simple after deletion of x. As already seen we have $S(x) = \{1, 3, 5, 7\}$, we also have $\{x, 2, 8\} \subset S(1)$, $\{1, 3\} \subset S(2)$... The restriction of the relation S inside the neighborhood of x is depicted Fig. 2 (d). We see that, if we consider a 2-cell as a closed path included in a unit cube, we have the structure of a 2D combinatorial manifold. In Fig. 2 (e), a configuration which could appear in a strong 18-surface is represented. We could expect that the restriction of the relation S inside the neighborhood of x be the one depicted

Fig. 2 (f): we see that the central point is surrounded by a simple closed curve $1, 2, 3, 4, 6, 7, 8$.

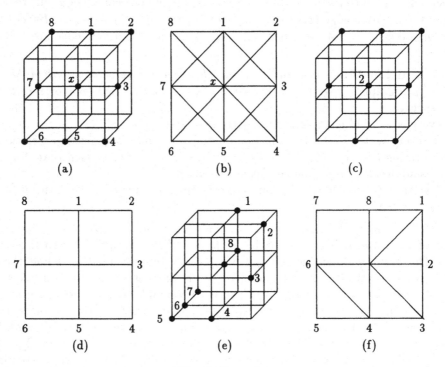

Fig. 2. Examples.

It follows the idea of considering the following graph for extracting a surface structure from a discrete surface in Z^3:

Definition 16: Let $X \subset E$. We define the graph $G_n(X)$ the vertices of which are the points of X and such that, for all x, y of X, x is adjacent to y if:

1) x and y are n-adjacent; and
2) x and y are not n-simple points for X; and
3) x is n-simple for $X \setminus \{y\}$; and
4) y is n-simple for $X \setminus \{x\}$.

Let G be a graph. A loop of G is *simple* if any vertex of the loop is adjacent to exactly two vertices in the loop.

Definition 17:

Let $X \subset E$. We define the set $F_n(X)$ composed of all simple loops for the graph $G_n(X)$ such that:

- for $n = 6$, these loops are included in a unit square;
- for $n = 6^+$, 18, 26, these loops are included in a unit cube.

The following theorem is the main result of this paper. It shows that the main existing notions of discrete surfaces in Z^3 are 2D combinatorial manifolds.

Theorem 18: *Let* $X \subset E$.

If X is a Morgenthaler n-surface, then $[G_n(X), F_n(X)]$ is a 2D combinatorial manifold, with $n = 6$ and $n = 26$.

If X is a Malgouyres 18-surface, then $[G_{18}(X), F_{18}(X)]$ is a 2D combinatorial manifold.

If X is a strong n-surface, then $[G_n(X), F_n(X)]$ is a 2D combinatorial manifold, with $n = 18$ and $n = 26$.

In Fig. 3 (a), a strong 26-surface X is depicted. The cavity of this surface is made of three points which constitute a "corner". The graph $G_{26}(X)$ is depicted Fig. 3 (b). It may be seen that $[G_{26}(X), F_{26}(X)]$ is a 2D combinatorial manifold.

6 Proof of the theorem

The proof of Th. 18 has been made with the help of a computer. Even with a computer this proof is not obvious. The reason is that proposed definition of combinatorial manifold involves the checking of the 125-neighborhood of a point: see the above discussion for the configuration of Fig. 2 (a) where some assumptions about the 125-neighborhood of the central point has been made for recovering the structural description of Fig. 2 (d). An exhaustive checking of all the 2^{125} configurations in this neighborhood is out of the reach of computers. This explains that we have to establish some intermediate lemmas in order to induce the properties of combinatorial manifolds from the 26-neighborhood of a point.

First of all, we introduce the notion of extensible configurations (see also [17]). Let us consider the configuration of Fig. 1 (e). It satisfies the conditions for Morgenthaler's 26-surface points (Def. 5). Nevertheless, it is impossible that such a configuration appear in a Morgenthaler's 26-surface. The reason is that this configuration is not extensible: it is not possible that all the points 26-adjacent to the central point could satisfy the conditions of Morgenthaler's 26-surface points. We give a precise definition of these cases:

Let $x \in E$. A *configuration of x* is a subset of $N_{26}(x)$ which contains x. Let $\mathcal{K}_x$ be a set of configurations of x. We say that $C_x \in \mathcal{K}_x$ is *extensible* if, for each point y of C_x there exists a configuration $C'_x \in \mathcal{K}_x$ such that C'_y, which is the translation of C'_x by the vector xy, satisfies $C'_y \cap N_{26}(x) = C_x \cap N_{26}(y)$.

We present now the way for proving Th. 18. By Th. 12 and 13, we have only to prove it for Morgenthaler's 6-surfaces and for strong n-surfaces ($n = 18, 26$).

6.1 Morgenthaler's 6-surfaces

First, a list of all possible configurations which satisfy the conditions for Morgenthaler's 6-surfaces has been made. Such an exhaustive checking is within the reach of computers since it involves only 2^{26} cases. Second all non-extensible configurations have been eliminated. Then, the following lemma was proved by checking all extensible configurations, (xG_ny means that x and y are adjacent for the graph $G_n(X)$):

Lemma 19: *Let X be a Morgenthaler's 6-surface.*
If $x \in X$ and $y \in X$ are 6-neighbors, then xG_6y.

On the other hand, by Def. 16, if xG_6y, then x and y are necessarily 6-neighbors. Thus it is possible to prove the theorem by checking again all extensible configurations: with Lemma 19, it is possible to recover the graph $G_6(X)$ and the set of loops $F_6(X)$ which appear in the neighborhood of the central point. The conditions for 2D combinatorial manifolds were verified.

6.2 Strong 26-surfaces

We see that the characterization of Th. 11 for strong n-surfaces is not "fully local": the knowledge of $|X|^x$ is not sufficient to decide if x satisfies the four properties. For checking the characterization, we need to know $|X|^x$ but we also need to know the distribution of the points of $|\overline{X}|^x$ between $|A|^x$ and $|B|^x$. In fact, since the symmetry of the four conditions with respect to A and B, we see that it is sufficient to know this distribution up to a renaming of A and B. More precisely, it is sufficient to know, for each x of X, a labeling of $|\overline{X}|^x$:

Definition 20: Let $X \subset E$ be an n-separating set and let $x \in X$. A *labeling* of $|\overline{X}|^x$ is a map $f_x \colon |\overline{X}|^x \longrightarrow \{0, 1\}$ such that $\{f_x^{-1}(0), f_x^{-1}(1)\} = \{|A|^x, |B|^x\}$.

The knowledge of a labeling is necessary only if there is a component of $|\overline{X}|^x$ not $\overline{n}$-adjacent to x, (see, for example, the configuration depicted Fig. 1 (g)). The following lemmas allow to characterize these cases (see [6] for the proof).

Lemma 21:
Let X be a strong 26-surface and let $x \in X$. If $|\overline{X}|^x$ contains a 6-component not 6-adjacent to x, then this 6-component is composed solely of one point. Furthermore this point is necessarily strictly 26-adjacent to x.

Lemma 22:
Let X be a strong 26-surface. Let $x \in X$ and let $y \in N_{18}^+(x) \cap X$. If $N_6^*(x) \cap N_6^*(y)$ is a subset of the same 6-component of $|\overline{X}|^x$, then y is 6-adjacent to a one-point component of $|\overline{X}|^x$. Furthermore this one-point component and $N_6^*(x) \cap N_6^*(y)$ will belong to two different 6-components of $\overline{X}$.

For proving Th. 18, we first make a list of all possible configurations such that there exists a labeling for which the conditions of Th. 11 are satisfied. We eliminate all non-extensible configurations. Then, we prove the following lemma by an exhaustive checking of all remaining configurations:

Lemma 23: *Let X be a strong 26-surface and let $x \in X$. We have:*
- *$\forall y \in N_6^*(x) \cap X$, $xG_{26}y$; and*
- *$\forall y \in N_{18}^+(x) \cap X$, if one of the two common 6-neighbors of x and y belong to X, then we do not have $xG_{26}y$; and*
- *$\forall y \in N_{18}^+(x) \cap X$, if the two common 6-neighbors of x and y belong to two different 6-components of $\overline{X}$, then $xG_{26}y$; and*
- *$\forall y \in N_{18}^+(x) \cap X$, if the two common 6-neighbors of x and y belong to the same 6-component of $\overline{X}$, then we do not have $xG_{26}y$.*

On the other hand, by the definition of simple points, if $y \in N_{26}^+(x) \cap X$, we do not have $xG_{26}y$. Thus, as for Morgenthaler's 6-surfaces, it is possible to examine all extensible configurations and, with Lemmas 22 and 23, to recover the graph

$G_{26}(X)$ and the set of loops $F_{26}(X)$ which appear in the neighborhood of the central point. The conditions for 2D combinatorial manifolds were verified.

6.3 Strong 18-surfaces

As for strong 26-surfaces, we first make a list of all possible configurations such that there exists a labeling for which the conditions of Th. 11 are satisfied. We eliminate all non-extensible configurations. We then prove the following lemmas by an exhaustive checking of all these configurations:

Lemma 24:
If X is a strong 18-surface, then $\forall x \in X$, $|\overline{X}|^x$ admits an unique labeling.

Lemma 25: *Let X be a strong 18-surface and let x and y be two points of X. Let C be any back-component of X, i.e. $C = A$ or $C = B$. We say that a 6-path π from x to y is a C-path if all points of π, except x and y, belong to C. We say that x and y are C^k-connected if there is a C-path from x to y and if the minimal length of a C-path from x to y is equal to k.*
We have:
- $\forall y \in N_6^*(x) \cap X$, $xG_{18}y$; and
- $\forall y \in N_{18}^+(x) \cap X$, $xG_{18}y$ if and only if x and y are A^k-connected and B^l-connected, with $k + l \leq 6$; and
- $\forall y \in N_{18}^+(x) \cap X$, if the two common 6-neighbors of x and y belong to $\overline{X}$, then we have $xG_{18}y$ if and only if the two common 6-neighbors belong to two different 6-components of $\overline{X}$.

The problem with strong 18-surfaces is that, if a point y is such that $y \in N_{18}^+(x) \cap X$, and if only one of the two common 6-neighbors of x and y belong to X, then we may have $xG_{18}y$ but it is also possible that x and y are not adjacent under the G_{18} relation. It follows that it is not always possible to recover the G_{18} graph by examining the 26-neighborhood of a point. For these cases, we make two assumptions: we first suppose that $xG_{18}y$ and then we suppose that x and y are not adjacent under the G_{18} relation. With this exhaustive checking and with Lemmas 24 and 25, Th. 18 was proved.

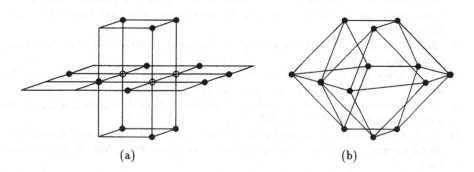

(a)   (b)

Fig. 3. A strong 26-surface X and the graph $G_{26}(X)$.

References

1. E. Artzy, G. Frieder, and G.T. Herman, "The theory, design, implementation and evaluation of a three-dimensional surface detection algorithm", *Comp. Graphics and Im. Proc.*, Vol. 15, pp. 1-24, 1981.
2. G. Bertrand, "Simple points, topological numbers and geodesic neighborhoods in cubic grids", *Pattern Rec. Letters*, Vol. 15, pp. 1003-1011, 1994.
3. G. Bertrand, "On P-simple points", *C.R. Académie des Sciences*, Série I, t. 321, p.1077-1084, 1995.
4. G. Bertrand, "P-simple points and 3D parallel thinning algorithms", submitted for publication.
5. G. Bertrand and R. Malgouyres, "Topological properties of discrete surfaces", *Lect. Notes in Comp. Sciences*, 1176, Springer Verlag, pp. 325-336, 1996.
6. G. Bertrand and R. Malgouyres, "Local property of strong surfaces", *SPIE Vision Geometry VI.*, to appear, 1997.
7. J. Françon, "Discrete combinatorial surfaces", *CVGIP: Graphical Models and Image Processing*, Vol. 57, pp. 20-26, 1995.
8. G.T. Herman, "Discrete multidimensional surfaces", *CVGIP: Graph. Models and Image Proc.*, Vol. 54, 6, pp. 507-515, 1992.
9. Y. Kenmochi, A. Imiya, and N. Ezquerra, "Polyhedra generation from lattice points", *Lect. Notes in Comp. Sciences*, 1176, pp. 127-138, 1996.
10. T. Y. Kong, "A digital fundamental group", *Computer Graphics*, 13, pp. 159-166, 1989.
11. T.Y. Kong, "On the problem of determining whether a parallel reduction operator for n-dimensional binary images always preserves topology", *SPIE Vision Geometry II*, Vol. 2060, 69-77, 1993.
12. T.Y. Kong and A. Rosenfeld, "Digital topology: introduction and survey", *Comp. Vision, Graphics and Image Proc.*, 48, pp. 357-393, 1989.
13. R. Kopperman, P.R. Meyer and R.G. Wilson, "A Jordan surface theorem for three-dimensional digital spaces", *Discrete Comput. Geom.*, 6, pp. 155-161, 1991.
14. J.-O. Lachaud, "Topologically defined iso-surfaces", *Lect. Notes in Comp. Sciences*, 1176, pp. 245-256, 1996.
15. A. McAndrew and C. Osborne, "Algebraic methods for multidimensional digital topology", *SPIE Vision Geometry II*, Vol. 2060, pp. 14-25, 1993.
16. G. Malandain, G. Bertrand and N. Ayache, "Topological segmentation of discrete surfaces", *Int. Journal of Comp. Vision*, Vol. 10, 2, 183-197, 1993.
17. R. Malgouyres, "A definition of surfaces of Z^3", *Conf. on discrete Geometry for Comp. Imag.*, pp. 23-34, 1993. See also *Doctoral dissertation*, Université d'Auvergne, France, 1994.
18. R. Malgouyres, "About surfaces in Z^3", *5th Conf. on Discrete Geom. for Comp. Imag.*, pp.243-248, 1995.
19. R. Malgouyres and G. Bertrand, "Complete local characterization of strong 26-surfaces", *Int. Work. on Par. Im. An.*, to appear, 1997.
20. D.G. Morgenthaler and A. Rosenfeld, "Surfaces in three-dimensional images", *Information and Control*, 51, 227-247, 1981.
21. G.M. Reed, "On the characterization of simple closed surfaces in three-dimensional digital images", *Computer Vision, Graphics and Image Proc.*, Vol. 25, pp. 226-235, 1984.
22. G.M. Reed and A. Rosenfeld, "Recognition of surfaces in three-dimensional digital images", *Information and Control*, Vol. 53, pp. 108-120, 1982.

A Linear Algorithm for Constructing the Polygon Adjacency Relation in Iso-Surfaces of 3D Images

Serge Miguet

Laboratoire ERIC

Bât. L, Univ. Lyon II

5 av. P. Mendès-France

69676 Bron (France)

miguet@univ-lyon2.fr

Jean-Marc Nicod*

LIB

IUT Belfort-Montbeliard

BP 527 Rue Engel Gros

90016 Belfort Cedex

Jean-Marc.Nicod@

iut-bm.univ-fcomte.fr

David Sarrut

Laboratoire ERIC

Bât. L, Univ. Lyon II

5 av. P. Mendès-France

69676 Bron (France)

dsarrut@univ-lyon2.fr

Abstract

This paper proposes an optimal algorithm for constructing the surface adjacency relation in a list of polygons extracted from 3D medical images. The discrete nature of data allows us to build this adjacency relation in a time proportional to the number of triangles $\mathcal{T}$. We have payed a special attention on the memory requirements, since our method takes as input the surface extracted by the Marching-Cubes algorithm and does not make reference to the initial 3D dataset. Moreover, no additional temporary storage is needed to compute the relation.

1 Introduction

Since the development of 3D medical scanning devices, a tremendous number of techniques have been proposed for reconstructing, processing and visualizing the anatomical data. One of the most used approach for understanding the 3D structure of objects consists in extracting iso-surfaces from the volume. The geometric description of these data can then be visualized by the help of classical rendering algorithms, using various lighting and shading models.

The Marching-Cubes algorithm (MC) is probably the most popular of these surface-based techniques, and is widely used in medical imaging as well as in

---

*This research has been done while the author was member of ERIC, Université Lyon-2

other application domains (geoscience, biology, molecular systems, etc). The input dataset consists in a 3D regular mesh whose elements (the voxels) have gray-level values depending on the acquisition device (MRI, CT, etc). In its original version [LC87], the algorithm generates a list of polygons (triangles) corresponding to a threshold value. The extracted surface separates the voxels having value above and below this threshold. Despite this list being sufficient for visualization, some processing for geometric modeling, topological considerations or image analysis need more information such as connected surface components, surface orientation or curvature, inner volume, total area, and so on. All of these kernels need to access in a simple way to the neighborhood of a given polygon. Although this adjacency relation could be computed by the MC itself, it would need an important amount of additional memory and computing time.

The main goal of this paper is to propose an efficient algorithm for constructing the surface adjacency relation, taking as input the original extracted surface, without using the initial 3D dataset. We show that our algorithm has a linear (optimal) complexity in the number of polygons of the surface, without requiring additional memory. Experimental results on medical images containing more than 1 million triangles fully demonstrate the efficiency of our approach, since the time for constructing the adjacency relation is less than 10 seconds.

2 Statement of the problem

The principle of the MC algorithm is to consider a cell made by 8 input samples. If the iso-surface intersects this cell, we build this intersection with triangles. The whole surface is made by moving this cell in the image successively along the x, y, and z axis (see a part of such a surface on figure 3).

The data generated by MC consists of two lists: a list of vertices and a list of triangles (solid tables in figure 1). Each vertex is represented by three position coordinates and three coordinates of the vector normal to the surface . One triangle is a triple of integers corresponding to the three vertex indices in the first list. This allows to store the surface in a compact and non-redundant way [MN95, ZN94]. Our goal is to build the surface adjacency relation $\mathcal{R}$ such that two distinct triangles t and t' are in relation by $\mathcal{R}$ if and only if they have an edge in common. We use a slightly modified version of MC as proposed in [Lac96], that has been proved to generate simple, oriented and closed surfaces. Thanks to these properties, an edge is shared by two and only two faces, which implies that a given triangle of the surface has exactly three neighbors. The adjacency relation will thus be stored by adding to each triangle the indices of its three neighbors (grayed table in figure 1).

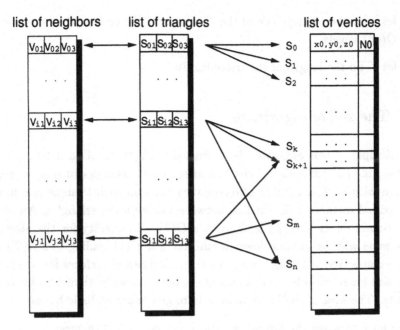

Figure 1: Output data structure of the MC algorithm

3 Algorithms for adjacency construction

In this section, we introduce two different adjacency construction algorithms that require no additional memory. Indeed, it would be possible to use a 3D data structure or an additional list associated with vertices to store the relation, but the important needs of memory make these solutions practically unusable with real medical images.

The first algorithm we present uses an intuitive approach, but its complexity makes it unsuitable for surfaces with a lot of triangles. We then explain our modified algorithm, that is proved to be optimal in section 3.3.

3.1 Notations

- let t be a triangle of the surface,

- let $C(t) = (i, j, k)$ be the integer coordinates of the cell containing t,

- let $\mathcal{I}(t)$ be position of t in the triangles list,

- let ω be the maximum number of triangles that might be generated in one cell. $\omega = 6$ in the version of MC algorithm we use.

- let n be the average side of the 3D dataset. We assume thus that we have $O(n^3)$ voxels.

- let $\mathcal{T}$ be the length of the triangles list.

3.2 The naive algorithm

The principle of this algorithm is to find the neighbors of each triangle t, by searching in the remainder of the list, three other triangles having an edge in common with t. The number of operations done on each triangle is related to the maximal distance Δ in the list, between two adjacent triangles. An obvious upper bound on Δ is $O(\mathcal{T})$ leading to a $O(\mathcal{T}^2)$ complexity for the algorithm. In the same way, an obvious lower bound on Δ is $O(1)$ leading to a $\Omega(\mathcal{T})$ lower bound complexity. Although, we can exhibit families of surfaces for which these upper and lower bounds are reached, these bounds can be tightened for real life datasets. The idea of the improvement is to give more realistic bounds on Δ.

We have to make the following hypotheses on our 3D images:

H1: the number of surface elements is a quadratic function of n, i.e. $\mathcal{T} = \Theta(n^2)$.

H2: the number $\mathcal{T}_k$ of surface elements generated in each layer k is bounded by a linear function of n, i.e. $\mathcal{T}_k = O(n)$.

These hypotheses exclude mathematically generated families of images, such as fractals that can generate $\Theta(n^3)$ surface elements, or wireframe objects that can lead to $O(n)$ surface elements only.

Theorem 1

> *The naive algorithm has a $O(\mathcal{T}^{\frac{3}{2}})$ complexity on images verifying hypotheses H1 and H2.*

Proof:

- from H1, we have: $\mathcal{T} = \Theta(n^2) \Rightarrow \exists \alpha > 0 \mid \alpha n^2 \leq \mathcal{T}$, thus

$$n \leq \sqrt{\frac{\mathcal{T}}{\alpha}} \qquad (1)$$

- from H2, we have:

$$\mathcal{T}_k = O(n) \Rightarrow \exists \beta > 0 \mid \mathcal{T}_k \leq \beta n, \qquad (2)$$

- let t be a triangle generated on layer k. The number of triangles we have to scan to find the neighbors of t can be bounded by the number of triangles generated on both layers k and $k+1$:

$$\Delta \leq \mathcal{T}_k + \mathcal{T}_{k+1}$$

from equations (1) and (2), we have:

$$\Delta \leq 2\beta\sqrt{\frac{\mathcal{T}}{\alpha}},$$

thus $\Delta = O(\mathcal{T}^{\frac{1}{2}})$.

An upper bound on the complexity of the naive algorithm is thus $O(\Delta \times \mathcal{T}) = O(\mathcal{T}^{\frac{3}{2}})$. □

Nevertheless, this complexity is still too high to use this algorithm for a surface coming from 3D medical images ($\mathcal{T} > 10^6$). So we have to propose another way to construct the adjacency.

3.3 A linear adjacency construction algorithm

The idea is now to avoid the entire scan of the triangle list from the current position. We propose to start the scan from different positions defined by the previous searches. Let t be a triangle of the surface belonging to cell $\mathcal{C}(t) = (i, j, k)$. The neighbors t' of t that are still unknown can be found in four different cells only, that are $\mathcal{C}_I = \mathcal{C}(t)$ itself or the three 6–neighboring cells of $\mathcal{C}(t)$ that have not yet been scanned, i.e. $\mathcal{C}_X = (i+1, j, k)$, $\mathcal{C}_Y = (i, j+1, k)$ and $\mathcal{C}_Z = (i, j, k+1)$. The corresponding subscript I, X, Y or Z is called the *direction* of the neighbor t' and is noted $\mathcal{D}$ in the following. This direction can easily be determined from the coordinates of the edge common to t and t'. Figure 2 illustrates these four cases. An additional fifth case noted $\mathcal{D} = \emptyset$ will be used in section 4 when the triangle has already been scanned by the algorithm, because it was located before t in the list.

In the case where $\mathcal{D} = I$ or $\mathcal{D} = X$, the neighbor t' is known to be close to t in the list and can be searched exhaustively in a constant time. It is only if $\mathcal{D} = Y$ or $\mathcal{D} = Z$, that we use an optimization.

Even if we do not know $\mathcal{I}(t')$, we know the exact position $\mathcal{I}(u')$ of the last triangle we searched for (from a given triangle u) in direction $\mathcal{D}$. The main idea of our algorithm consists in starting the search of t' in a neighborhood of u'.

The algorithm is justified by the following result:

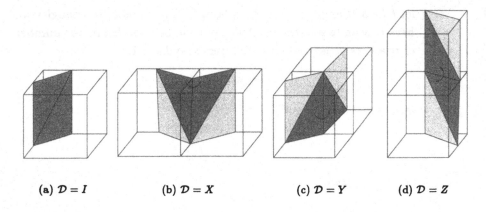

(a) $\mathcal{D} = I$ (b) $\mathcal{D} = X$ (c) $\mathcal{D} = Y$ (d) $\mathcal{D} = Z$

Figure 2: the 4 configurations corresponding to the 4 values of $\mathcal{D}$.

Theorem 2

> Let t be the current triangle in the list known to have a neighbor t' in direction $\mathcal{D}$ ($\mathcal{D} = Y$ or $\mathcal{D} = Z$), and let u be the last scanned triangle having had a neighbor u' in the same direction $\mathcal{D}$. The following property holds:
>
> $$\mathcal{I}(t') > \mathcal{I}(u') - \omega$$

Proof:

The different cells (i, j, k) are implicitly ordered by the scanning performed by MC. Following this order, the number of cells separating $\mathcal{C}(u)$ and $\mathcal{C}(t)$ is equal to the number of cells separating $\mathcal{C}(u')$ and $\mathcal{C}(t')$.

- If u and t have been generated in the same cell (i.e. $\mathcal{C}(u) = \mathcal{C}(t)$), then u' and t' are also generated in the same cell (i.e. $\mathcal{C}(u') = \mathcal{C}(t')$). We do not know if u' is generated before t', but since there are at most ω triangles per cell, it is sufficient to start the search of t' from index $\mathcal{I}(u') - \omega + 1$.

- Else (i.e. $\mathcal{C}(u) \neq \mathcal{C}(t)$), the fact that u is before t in the list implies that $\mathcal{C}(u)$ was scanned before $\mathcal{C}(t)$ by MC. We deduce that $\mathcal{C}(u')$ and $\mathcal{C}(t')$ were also scanned in that order which allows to conclude that $\mathcal{I}(t') > \mathcal{I}(u')$.

In these two cases, it is easy to check that the announced result is verified.

$\square$

The only data structure needed to control this adjacency construction algorithm is reduced to three addresses in the triangles list: the current index $\mathcal{I}(t)$, and the last values $\mathcal{I}_Y$ and $\mathcal{I}_Z$ of $\mathcal{I}(u')$, corresponding respectively to $\mathcal{D} = Y$ and $\mathcal{D} = Z$.

The figure 4 represents a part of the triangles list generated from the example shown in figure 3. In this case, the triangle t has one of its neighbors, named t', in the direction $\mathcal{D} = Y$. Before storing the relation between t and t', $\mathcal{I}_Y$ was equal to $\mathcal{I}(u')$ since (u, u') was the last couple of triangles in direction $\mathcal{D} = Y$. t' could thus be searched from the position of u' instead of the one of t as in the naive algorithm. After having established the relation between t and t', the new value of $\mathcal{I}_Y$ is $\mathcal{I}(t')$.

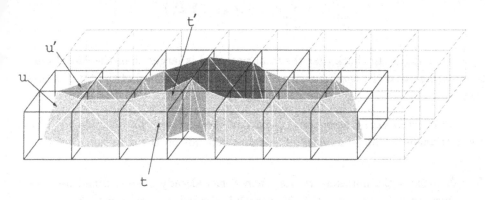

Figure 3: An example of a generated surface in a layer.

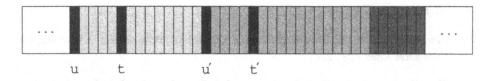

Figure 4: A example of a triangle list generated from the previous example

4 Complexity analysis

In this section, we present a proof of the linearity of the algorithm we have presented. The algorithm scans the whole list of triangles, and searches for the neighbors of the current one. Although some of these searches are costly (and can be as expensive as in the naive algorithm), we will show that they occur seldom enough to be amortized by many constant cost of other searches. This is a typical proof using the amortized analysis. We use the potential method of D. D. Sleator, with the notations of [CLR90]:

Let D_i be the data structure at the ith step of the algorithm. This method associates to the data structure a potential function Φ who maps each data structure D_i to a real number $\Phi(D_i)$. Let c_i be the actual cost of the ith operation. By definition, the amortized cost $\hat{c}_i$ of the ith operation is : $\hat{c}_i = c_i + \Phi(D_i) - \Phi(D_{i-1})$.

The main property used in amortized analysis is that the total amortized cost after n steps is equal to the total actual cost of these n steps plus the increase of potential:

$$\sum_{i=1}^{n} \hat{c}_i = \sum_{i=1}^{n} c_i + \Phi(D_n) - \Phi(D_0) \tag{3}$$

It is sufficient to check that the final potential is larger than the initial one, to guarantee that the total amortized cost is an upper bound of the total actual cost.

The ith step of the algorithm consists in the search of one neighbor t' of a given triangle t. The actual cost of searches depends on the direction $\mathcal{D}$ of t':

$\mathcal{D} = \emptyset$:   this is the notation we use when t' has already been scanned because located before t in the list. No search is needed, and we can let $c_i = 1$.

$\mathcal{D} = I$: t and t' are in the same cell, with distant of at most $\omega - 1$ triangles. We have thus $c_i < \omega$.

$\mathcal{D} = X$: t' is in the cell following t, with distant of at most $2\omega - 1$ triangles. We have thus $c_i < 2\omega$.

$\mathcal{D} = Y$:   the search for t' starts from index $\mathcal{I}_Y$. Let Δ_Y be the number of scanned triangles until t' is reached. We have thus $c_i = \Delta_Y$.

$\mathcal{D} = Z$:   the search for t' starts from index $\mathcal{I}_Z$. Let Δ_Z be the number of scanned triangles until t' is reached. We have thus $c_i = \Delta_Z$.

We are now ready to announce the main result of the paper:

Theorem 3

> *The algorithm described in section 3.3 has a linear complexity in the number of triangles*

Proof:

The expression of Φ we use for the potential method is:

$$\Phi(D_i) = i - \mathcal{I}_Y - \mathcal{I}_Z \qquad (4)$$

It is easy to check that this potential is null at the beginning of the algorithm, and positive at the end. Indeed, there are $\mathcal{T}$ triangles having each 3 edges. The last value of i is thus $3\mathcal{T}$, and $\mathcal{I}_Y$ and $\mathcal{I}_Z$ are smaller than $\mathcal{T}$.

Let us now compute the increase of potential associated to the five possible values for $\mathcal{D}$:

$\mathcal{D} = \emptyset$, $\mathcal{D} = I$ or $\mathcal{D} = X$:   in these three cases, $\mathcal{I}_Y$ and $\mathcal{I}_Z$ are not modified, thus $\Phi(D_i) - \Phi(D_{i-1}) = 1$,

$\mathcal{D} = Y$:   the new value of $\mathcal{I}_Y$ corresponds to the position of t'. It has thus been increased of Δ_Y, so $\Phi(D_i) - \Phi(D_{i-1}) = 1 - \Delta_Y$,

$\mathcal{D} = Z$:   the new value of $\mathcal{I}_Z$ corresponds to the position of t'. It has thus been increased of Δ_Z, so $\Phi(D_i) - \Phi(D_{i-1}) = 1 - \Delta_Z$,

By adding the actual cost with this increase of potential, we check that the amortized cost of the each of the five operations is $O(1)$. We can thus deduce that the $3\mathcal{T}$ operations are done in complexity $O(\mathcal{T})$.

$\square$

Since any algorithm for this problem has to visit at least once each triangle, our approach is optimal.

5   Experiments and results

In this section, we show experiments on the implementation of our two approaches for constructing the polygon adjacency relation.

In figure 5, we test the two algorithms on several surfaces extracted from 3D medical images. The computation time, for both figures, is obtained with the same set of surfaces. We can observe experimentally the linearity of our approach. For the surface made with 1.5×10^6 triangles, the time decreases from 3000 seconds to 10 seconds.

Figure 6, shows graphically the benefits of our improvement: we plot the cumulated cost of searches as a function of the displacement Δ, performed to

134

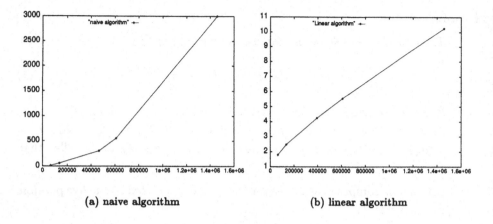

(a) naive algorithm (b) linear algorithm

Figure 5: Comparison of the computation time between two different algorithms for constructing the polygon adjacency relation.

find the neighbor. The surface of the histograms correspond to the cost of the algorithm. We clearly observe three modes in the first histogram 6(a) corresponding to the three directions of searches $\mathcal{D} = I, X, \mathcal{D} = Y$ or $\mathcal{D} = Z$. These modes completely disappear in figure 6(b) meaning that most of the searches are completely local.

6 Conclusion

We have introduced an optimal algorithm for constructing the adjacency relation in a triangles list computed by the Marching-Cubes algorithm. The main advantage of this technique is to require no additional memory, which is a crucial point in 3D medical imaging. We have used amortized analysis to prove the optimal complexity of our algorithm. Numerical experiments demonstrate practically the efficiency of the approach.

A parallel version of this algorithm has been developed and integrated in our parallel 3D medical imaging environment [Mig95]. It has been chained with a parallel connected components algorithm, adapted from Perroton [Per94]. This will make it possible to interactively select connected objects from a 3D visualization, or to initiate a segmentation algorithm based on active surfaces as proposed in [LB94].

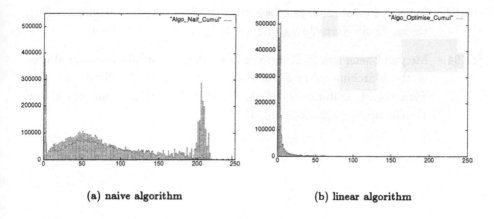

(a) naive algorithm   (b) linear algorithm

Figure 6: Distribution of the cumulative cost for each different displacement in the list.

References

[CLR90] Thomas Cormen, Charles Leiserson, and Ronald Rivest. *Introduction to Algorithms*. MIT Press, 1990.

[Lac96] Jacques-Olivier Lachaud. Topologically defined isosurfaces. In Serge Miguet, Annie Montanvert, and Stéphane Ubéda, editors, *Discrete Geometry for Computer Imagery*, volume 1176 of *Lecture Notes in Computer Science*, pages 245–256. Springer, November 1996.

[LB94] Jacques-Olivier Lachaud and Eric Bainville. A discrete model following topological modifications of volumes. In Jean-Marc Chassery and Annick Montanvert, editors, *Discrete geometry for computer imagery*, pages 183–194, Grenoble (France), September 1994.

[LC87] William E. Lorensen and Harvey E. Cline. Marching cubes : a high resolution 3D surface construction algorithm. *Computer Graphics*, 21(4):163–169, July 1987.

[Mig95] Serge Miguet. Un environnement parallèle pour l'imagerie 3D. Mémoire d'habilitation à diriger des recherches, Laboratoire de l'informatique du Parallélisme, 46 allée d'Italie, 69364 Lyon Cedex 07, December 1995.

[MN95] Serge Miguet and Jean-Marc Nicod. An optimal parallel iso-surface extraction algorithm. In *Fourth International Workshop on Parallel Image Analysis (IWPIA'95)*, pages 65–78. Laboratoire de l'Informatique du Parallélisme, ENS Lyon (France), December 1995.

136

[Per94] Laurent Perroton. *Segmentation Parallèle d'Images Volumiques.* PhD thesis, Ecole Normale Supérieure de Lyon, December 1994.

[ZN94] Meiyun Zheng and H.T. Nguyen. An Efficient Parallel Implementation of the Marching-cubes Algorithm. In L. Decker, W. Smit, and J.C. Zuidervaart, editors, *Massively Parallel Processing Applications and Development*, pages 903–910, 1994.

Topology

Digital Lighting Functions

R. Ayala[1], E. Domínguez[2], A.R. Francés[2], A. Quintero[1]

[1] Dpt. de Algebra, Computación, Geometría y Topología. Facultad de Matemáticas. Universidad de Sevilla. Apto. 1160. E-41080 – Sevilla. Spain.
e-mail: quintero@cica.es
[2] Dpt. de Informática e Ingeniería de Sistemas. Facultad de Ciencias. Universidad de Zaragoza. E-50009 – Zaragoza. Spain.
e-mail: ccia@posta.unizar.es

Abstract. In this paper a notion of lighting function is introduced as an axiomatized formalization of the "face membership rules" suggested by Kovalevsky. These functions are defined in the context of the framework for digital topology previously developed by the authors. This enlarged framework provides the (α, β)-connectedness $(\alpha, \beta \in \{6, 18, 26\})$ defined on $\mathbb{Z}^3$ within the graph-based approach to digital topology. Furthermore, the Kong-Roscoe (α, β)-surfaces, with $(\alpha, \beta) \neq (6, 6), (18, 6)$, are also found as particular cases of a more general notion of digital surface.

Keywords: Lighting function, digital surface, pixel connectivity, digital topology.

1 Introduction

In [1,2] we introduced a framework for digital topology whose main feature is to provide a link between digital spaces and Euclidean spaces. This framework consists of a multilevel architecture made up of five levels each of them representing a different level of abstraction for a digital picture, increasing from its digital structure to the continuous perception that an observer takes on it.

The starting level is a polyhedral complex, called the device level, which represents the physical layout of the pixels in the digital space, and so the neighbouring relationship considered among them. This relationship is abstracted by means of a graph, called the logical level. Two further levels serve as a bridge towards an Euclidean polyhedron, where every digital picture is associated with a subpolyhedron called its continuous analogue.

With this framework one takes advantage of the knowledge from continuous topology to obtain results in digital topology, by translating, whenever it is possible, not only the statements but also the proofs of the corresponding continuous ones to the logical level. Indeed, this method has allowed us to introduce a general notion of digital n-manifold extending the Morgenthaler (26,6)-surfaces [1], and then to prove a generalized digital index theorem for these n-manifolds [2].

Another interesting aspect of this framework is that it gathers, at least partially, some of the various approaches to digital topology that have been appeared in literature, such as those of Kovalevsky [8], Khalimsky [5], and the graph-based spaces due to Rosenfeld and other authors [7,11,6].

Concerning the latter approach, only some of the most usual graph-based spaces, as the (8,4)- and (26,6)-connected spaces (and their generalization to arbitrary dimension) or the hexagonal one, were found as the logical level of some device level. So that, this framework was not general enough to deal with all the graph-based spaces. This is so because each device level determines a single neighbouring relationship on the pixels, and so the logical level is fixed. The goal of this paper is to present an improved version of that framework in order to avoid this restriction. This is done by adding what we call a lighting function to the architecture quoted above (see §2). This allows us, given a device level, to select the neighbouring relationship we want to work with. In this way, the ability for translating results from continuous topology is preserved, and still the (α, β)-connectedness can be defined in this setting for all pairs (α, β) with $\alpha, \beta \in \{6, 18, 26\}$ (see §4). Furthermore, the (α, β)-surfaces, for $(\alpha, \beta) \neq (6,6), (18,6)$, are also found as particular cases of a more general notion of digital surface (see §5). Finally, it is worth pointing out that this new version provides a single digital notion of connectedness which works for both the digital object and its complement (see §3).

We refer to [12] for all notions in polyhedral topology contained in this paper. For recent trends in digital topology see [4].

2 Lighting functions and digital spaces

As in [1,2], a digital space consists of a multilevel architecture which provides a bridge for transferring definitions, statements and proofs from continuous topology to digital topology.

The first level of a digital space, called the *device level*, is used to represent the spatial layout of the pixels, which are represented by the n-cells of a homogeneously n-dimensional locally finite polyhedral complex K. Namely, K is a complex of convex cells (polytopes) such that each cell is face of a finite number (non-zero) of n-cells. If σ is a face of γ we shall write $\sigma \leq \gamma$. If $| K |$ denotes the underlying polyhedron of K, a centroid-map is a map $c : K \rightarrow | K |$ such that $c(\sigma)$ belongs to the interior of σ. The set of all n-cells of K will be denoted by $\mathrm{cell}_n(K)$. Given a device level K, a *digital object in K* is a subset of the set $\mathrm{cell}_n(K)$ of n-cells in K.

To avoid connectivity paradoxes, Kovalevsky points out in [8] the convenience of associating to each digital object some set of lower dimensional cells in K. These cells would indicate which pairs of n-cells should be considered adjacent. For this, Kovalevsky makes two proposals. On one hand, he suggests to encode a digital image by specifying not only what pixels (n-cells) are in the object but also the faces of these pixels which are associated to it. On the other hand, to save memory space, Kovalevsky observes that some global face membership rule can be used; that is, "a rule specifying the set membership of the faces of every n-cell as a function of the membership of the n-cell itself". We have adopted this last point of view, which has been formalized through the notion of lighting function. To introduce this notion we need the following notation.

Given a cell $\alpha \in K$ and a digital object $O \subseteq \text{cell}_n(K)$, the *star of α in O* is the set $\text{st}_n(\alpha; O) = \{\sigma \in O : \alpha \leq \sigma\}$, and the *support of O*, $\text{supp}(O)$, is the set of all cells $\alpha \in K$ such that $\alpha = \cap\{\sigma : \sigma \in \text{st}_n(\alpha; O)\}$. Observe that if $\text{st}_n(\alpha; O)$ has only one element, then $\alpha \in \text{supp}(O)$ if and only if $\alpha \in O$, and thus $\text{st}_n(\alpha; O) = \{\alpha\}$. To ease the writing, when the digital object is the whole set $\text{cell}_n(K)$ we shall write $\text{supp}(K)$ and $\text{st}_n(\alpha; K)$ instead of $\text{supp}(\text{cell}_n(K))$ and $\text{st}_n(\alpha; \text{cell}_n(K))$, respectively. Finally, we shall write $\mathcal{P}(A)$ for the family of all subsets of a given set A.

Definition 1. Given a complex K, a function $f : \mathcal{P}(\text{cell}_n(K)) \times K \to \{0, 1\}$ is said to be a *lighting function* on K if it verifies the following properties for all $O \in \mathcal{P}(\text{cell}_n(K))$ and $\alpha \in K$.

(F1) If $\alpha \notin \text{supp}(O)$ then $f(O, \alpha) = 0$.   (F3) $f(O, \alpha) = f(\text{st}_n(\alpha; O), \alpha)$.

(F2) If $\alpha \in O$ then $f(O, \alpha) = 1$.   (F4) $f(O, \alpha) \leq f(\text{cell}_n(K), \alpha)$.

In this way, given a digital object O, the complex K is partitioned into two subsets of cells. Namely, $\{\alpha \in K : f(O, \alpha) = 1\}$ which is associated to the object, and $\{\alpha \in K : f(O, \alpha) = 0\}$ associated to its complement. In addition, these properties formalize very natural and intuitive ideas. Property (F2) expresses that in order to display a digital object its pixels must be lighted, while (F1) says that the cells which are not the intersection of pixels of the object have nothing to do with its connectivity, and so we choose to get them dark. Property (F3) states that for a given object the lighting of a cell is a local property of the object; and finally, (F4) says that a cell $\alpha \in K$ is lighted for the global object $\text{cell}_n(K)$ whenever it is lighted for some small object $O \subseteq \text{cell}_n(K)$.

Given a lighting function f on K, the *logical level of a digital object O* is an undirected graph, $\mathcal{L}_O^f$, whose vertices are the centroids of n-cells in O and two of them $c(\sigma)$, $c(\tau)$ are adjacent if there exists a common face $\alpha \leq \sigma \cap \tau$ such that $f(O, \alpha) = 1$.

The *conceptual level of O* is the digraph $\mathcal{C}_O^f$ whose vertices are the centroids $c(\alpha)$ of all cells $\alpha \in K$ with $f(O, \alpha) = 1$, and its directed edges are $(c(\alpha), c(\beta))$ with $\alpha \leq \beta$.

The *simplicial analogue of O* is the order complex $\mathcal{A}_O^f$ associated to the digraph $\mathcal{C}_O^f$. That is, $\langle x_0, x_1, \ldots, x_m \rangle$ is an m-simplex of $\mathcal{A}_O^f$ if $x_0, x_1, \ldots, x_m$ is a directed path in $\mathcal{C}_O^f$. This simplicial complex defines the simplicial level for the object O in the architecture and, finally, the continuous level is represented by the underlying polyhedron $| \mathcal{A}_O^f |$ of $\mathcal{A}_O^f$. This polyhedron is called the *continuous analogue of O*.

Example 1. Every polyhedral complex $K \neq \emptyset$ admits the following lighting functions: $f_{\max}(O, \alpha) = 1$ if and only if $\alpha \in \text{supp}(O)$; $f_{\min}(O, \alpha) = 1$ if and only if $\alpha \in O$; and, $g(O, \alpha) = 1$ if and only if $\alpha \in \text{supp}(O)$ and $\text{st}_n(\alpha; K) \subseteq O$.

Notice that both $f_{\max}$ and g are distinct from $f_{\min}$ only if there exist two n-cells in K with a common face. On the other hand, if a cell $\alpha \in K$ is the intersection of a proper subset of n-cells in $\text{st}_n(\alpha; K)$ then $g \neq f_{\max}$. Moreover, each one of these lighting functions may induce different levels for a given digital

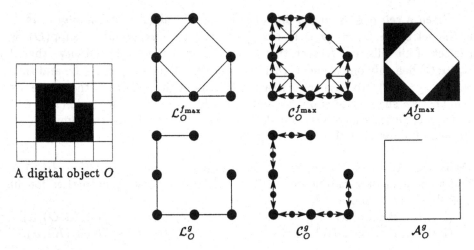

Fig. 1. Logical and conceptual levels, and simplicial analogues of a digital object for $f_{\max}$ and g in Example 1.

object, as Figure 1 shows for $f_{\max}$ and g. In this example, all levels for $f_{\min}$ are the same; namely, a set of seven discrete points. From this example it is evident that the levels for a digital object depend both on the complex K and on the lighting function f considered on it. So that, we define the notion of digital space as follows.

Definition 2. A *digital space* is a pair (K, f), where K is a homogeneously n-dimensional locally finite polyhedral complex and f a lighting function on K.

From now on, when working with a digital space (K, f) and if there is no place to confusion, for a digital object O we shall write $\mathcal{L}_O, \mathcal{C}_O, \mathcal{A}_O$ and $|\mathcal{A}_O|$ instead of $\mathcal{L}_O^f, \mathcal{C}_O^f, \mathcal{A}_O^f$ and $|\mathcal{A}_O^f|$ respectively to denote the corresponding levels of O. Moreover, if the digital object is the set $\mathrm{cell}_n(K)$ of all n-cells in K we shall write $\mathcal{L}_K, \mathcal{C}_K, \mathcal{A}_K$ and $|\mathcal{A}_K|$ for its levels, which will be called the levels of the whole digital space (K, f).

Next, we introduce some structural properties about digital spaces whose proofs are straightforward.

Proposition 3. *Let (K, f) be a digital space and $O \subseteq \mathrm{cell}_n(K)$ a digital object. Then (1) $\mathcal{L}_O$ is a (not necessarily full) subgraph of $\mathcal{L}_K$; (2) $\mathcal{C}_O$ is a full subgraph of $\mathcal{C}_K$; (3) $\mathcal{A}_O$ is a full subcomplex of $\mathcal{A}_K$.*

Clearly, any lighting function f on K is the characteristic function of some subset of $\mathcal{P}(\mathrm{cell}_n(K)) \times K$. However, not all the subsets of $\mathcal{P}(\mathrm{cell}_n(K)) \times K$ define a lighting function (for instance, if $K \neq \emptyset$, the characteristic function of the empty set does not verify the property (F2)).

Theorem 4. *The set of all lighting functions on a given complex K is a distributive complete lattice, whose greatest and least elements are $f_{\max}$ and $f_{\min}$, respectively. Moreover, it is a Boolean algebra if and only if $f_{\max} = g$.*

3 Connectedness in digital spaces

In this section we introduce the notion of connectedness for subsets of n-cells in a digital space. This notion includes, as particular cases, the connectedness for both digital objects and their complements. Afterwards, we shall prove that this notion of connectedness coincides with the corresponding topological notion in the continuous analogue.

Definition 5. Let O and O' be two disjoint digital objects in a digital space (K, f). Two distinct n-cells $\sigma, \tau \in O$ are said to be O'-*adjacent in* O if there exists a common face $\alpha \leq \sigma \cap \tau$ such that $f(O', \alpha) = 0$ and $f(O \cup O', \alpha) = 1$. An O'-*path in* O *from* σ *to* τ is a finite sequence $\{\sigma_i\}_{i=0}^m \subseteq O$ such that $\sigma_0 = \sigma$, $\sigma_m = \tau$ and σ_{i-1} is O'-adjacent in O to σ_i, for $i = 1, \ldots, m$.

The digital object O will be said O'-*connected* if for any pair of n-cells $\sigma, \tau \in O$ there exists an O'-path in O from σ to τ. An object $C \subseteq O$ is an O'-*component* of O if for any pair $\sigma, \tau \in C$ there exists an O'-path in O from σ to τ and none element in C is O'-adjacent in O to some element of $O - C$. Observe that any O'-component is O'-connected.

Given a digital object O in the digital space (K, f) the previous definitions provide an entire family of notions of connectedness for O in relation to another object O', when O' is allowed to range over the set of all subsets of $\text{cell}_n(K) - O$. The extreme cases, when $O' = \emptyset$ and $O' = \text{cell}_n(K) - O$, represent the connectedness of the digital object O itself and the connectedness of O as the complement of O', respectively.

Following this line, we will call *connected* to any object which is $\emptyset$-connected, and C is a *component* of O if it is a $\emptyset$-component. Moreover, it is easy to check that two n-cells $\sigma, \tau \in O$ are $\emptyset$-adjacent in O if and only if there exists $\alpha \leq \sigma \cap \tau$ such that $f(O, \alpha) = 1$. So that, $\sigma, \tau \in O$ are $\emptyset$-adjacent in O if and only if their centroids $c(\sigma)$ and $c(\tau)$ are adjacent as vertices of the logical level $\mathcal{L}_O$ of O. This justifies to call *adjacent in* O to any pair of n-cells which are $\emptyset$-adjacent in O, and then a *path in* O is just a $\emptyset$-path in O. These observations prove the following result.

Proposition 6. *A digital object O is connected if and only if its logical level $\mathcal{L}_O$ is a connected graph.*

Furthermore, when O is considered as the complement of the digital object $O' = \text{cell}_n(K) - O$, we get that $\sigma, \tau \in O$ are O'-adjacent in O if and only if there exists $\alpha \leq \sigma \cap \tau$ such that $f(O', \alpha) = 0$ and $f(\text{cell}_n(K), \alpha) = 1$. So that, two n-cells σ, τ are O'-adjacent in the complement of O' if and only if there exists $\alpha \in K$ whose centroid $c(\alpha)$ is adjacent to both $c(\sigma)$ and $c(\tau)$ in the complement $\mathcal{C}_K \setminus \mathcal{C}_{O'}$ of the conceptual level $\mathcal{C}_{O'}$ in $\mathcal{C}_K$. In this way, the connectedness of the complement of an object can be characterized in the conceptual level, but not in the logical level. Indeed, the complement of the object O shown in Figure 1 it is not O-connected in the digital space $(K, f_{\max})$, while the complement of the logical level $\mathcal{L}_O^{f_{\max}}$ of O in $\mathcal{L}_K^{f_{\max}}$ is connected.

Theorem 7 shows how these notions of connectedness are stated at each level of our architecture. This result is an immediate consequence of Theorem 8. Below, $L_1 \setminus L_2 = \{\alpha \in L_1 : \alpha \cap | L_2 | = \emptyset\}$ will stand for the *simplicial complement of L_2 in L_1*, where L_1 and L_2 are subcomplexes of a simplicial complex L.

Theorem 7. *Let O be a digital object. The following properties are equivalent: (1) O is connected. (2) $\mathcal{L}_O$ is a connected graph. (3) $\mathcal{C}_O$ is a connected digraph. (4) $\mathcal{A}_O$ is a connected simplicial complex. (5) $| \mathcal{A}_O |$ is a connected space.*

Moreover, if $O' = \mathrm{cell}_n(K) - O$, the following properties are equivalent: (1) O is O'-connected. (2) $\mathcal{C}_K \setminus \mathcal{C}_{O'}$ is a connected digraph. (3) $\mathcal{A}_K \setminus \mathcal{A}_{O'}$ is a connected simplicial complex. (4) $| \mathcal{A}_K | - | \mathcal{A}_{O'} |$ is a connected space.

Theorem 8. *Let O and O' be two disjoint digital objects in a digital space. The family $\mathcal{F}$ of O'-components of O can be described in any of the following ways*

(1) Conceptual level: $\mathcal{F} = \{O_G\}$, where $O_G = \{\sigma \in O : c(\sigma)$ is a vertex of $G\}$, and G ranges over the family of components of the digraph $\mathcal{C}_{O \cup O'} \setminus \mathcal{C}_{O'}$.

(2) Simplicial level: $\mathcal{F} = \{O_A\}$, where $O_A = \{\sigma \in O : c(\sigma) \in A\}$, and A ranges over the family of components of the simplicial complement $\mathcal{A}_{O \cup O'} \setminus \mathcal{A}_{O'}$.

(3) Continuous level: $\mathcal{F} = \{O_X\}$, where $O_X = \{\sigma \in O : c(\sigma) \in X\}$, and X ranges over the family of components of the space $| \mathcal{A}_{O \cup O'} | - | \mathcal{A}_{O'} |$.

We sketch a proof of this theorem. Firstly, the characterization in the conceptual level can be readily proved from the following proposition.

Proposition 9. (i) *Let $O_2 \subseteq O_1$ be two digital objects. If $c(\tau)$ is a vertex of the complement $\mathcal{C}_{O_1} \setminus \mathcal{C}_{O_2}$ then there exists an n-cell $\sigma \in O_1 - O_2$ such that $\tau \leq \sigma$.*

(ii) *Let O and O' be two disjoint digital objects in a digital space. Given two distinct n-cells $\sigma, \tau \in O$ there exists a O'-path in O from σ to τ if and only if their centroids $c(\sigma)$ and $c(\tau)$ are vertices of the same component of $\mathcal{C}_{O \cup O'} \setminus \mathcal{C}_{O'}$.*

Next, to obtain the characterization in the simplicial level from that in the conceptual level it is enough to observe that $\mathcal{C}_{O \cup O'} \setminus \mathcal{C}_{O'}$ can be identified with the 1-skeleton of $\mathcal{A}_{O \cup O'} \setminus \mathcal{A}_{O'}$. Finally, the characterization in the continuous level follows from the simplicial one and the next lemma from simplicial topology.

Lemma 10. *Let $L \subseteq K$ be a full subcomplex of a locally finite simplicial complex K. Then, the components of $| K | - | L |$ are in 1-1 correspondence with the components of $K \setminus L$.*

4 Lighting functions for the (α, β)-connectedness

Within the graph-based approach to digital spaces, due to Rosenfeld and other authors, many graphs on almost arbitrary grids of points have been used [7]. However, this section is only concerned with the most usual connectedness on the grid $\mathbb{Z}^3$, defined by means of the double adjacency (α, β), with $\alpha, \beta \in \{6, 18, 26\}$. The α-adjacency is considered to define the connection for digital objects and the β-adjacency for their complements. See [6] for a precise definition.

In this section we show how all types of (α, β)-connectedness on the grid $\mathbf{Z}^3$ can be recovered in our framework by selecting suitable lighting functions. These functions will be defined on the device level R^3, called the *standard cubical decomposition* of the 3-dimensional Euclidean space $\mathbb{R}^3$. That is, R^3 is the complex determined by the collection of unit 3-cubes in $\mathbb{R}^3$ whose edges are parallel to the coordinate axes and whose centres are in the set $\mathbf{Z}^3$. The centroid-map we will consider in R^3 associates to each cube σ its barycentre $c(\sigma)$. In particular, if $\dim \sigma = 3$ then $c(\sigma) \in \mathbf{Z}^3$, where $\dim \sigma$ stands for the dimension of σ. So that, every digital object O in R^3 can be identified with a subset of points in $\mathbf{Z}^3$. Henceforth we shall use this identification without further comment.

Definition 11. We say that a lighting function $f_{\alpha,\beta}$ on R^3 *provides the (α, β)-connectedness* if the two following properties hold for any digital object O: (1) O is connected if and only if it is α-connected; and, (2) whenever O is considered as the complement of the object $O' = \text{cell}_3(R^3) - O$, then O is O'-connected if and only if it is β-connected.

In Example 2 below we give some lighting functions providing the (α, β)-connectedness for each pair (α, β). For this we consider a new polyhedral decomposition of $\mathbb{R}^3$, denoted $R^3(\mathbf{Z}^3)$, consisting of unit cubes with vertices in $\mathbf{Z}^3$. To avoid misunderstandings, we keep the terminology "cube" for the 3-cells in R^3 and we call $\mathbf{Z}^3$-cell to the closed cubes in $R^3(\mathbf{Z}^3)$. Given a digital object O in R^3 and a $\mathbf{Z}^3$-cell C, the *configuration of O in C* is the set $C(O) = \{c(\sigma) \in \mathbf{Z}^3 : \sigma \in O\} \cap C$ of vertices of C which are the centroids of cells in O. Observe that the centre of C coincides with some 0-cell $\rho \in R^3$. So that, $C(O)$ is the set of centroids of cubes in $\text{st}_3(\rho; O)$. In Figure 2 are shown all the possible configurations of a given object after a suitable rotation or reflection.

Example 2. The lighting functions $f_{\alpha,\beta}^n$ listed below are providing the corresponding (α, β)-connectedness, for all pairs (α, β) with $\alpha, \beta \in \{6, 18, 26\}$. To prove this fact is a tedious but mechanical task, which involves to check properties (F1)-(F4) in Definition 1 and to prove that the components and the $(\text{cell}_3(R^3) - O)$-components of a given digital object O coincide with the α-components and the β-components, respectively.

a) $f_{6,6}^0(O, \alpha) = 1$ iff $\alpha \in O$ for $\dim \alpha = 3$, and $\alpha \in \text{supp}(O)$ for $\dim \alpha = 2$.

b) $f_{6,18}^0(O, \alpha) = 1$ iff $\alpha \in O$ for $\dim \alpha = 3$, $\alpha \in \text{supp}(O)$ for $\dim \alpha = 2$, $\text{st}_3(\alpha; R^3) \subseteq O$ for $\dim \alpha = 1$, and $\text{st}_3(\alpha; O)$ contains the configuration (6c) in Figure 2 for $\dim \alpha = 0$.

c) $f_{6,18}^1(O, \alpha) = f_{6,18}^0(O, \alpha)$ for $\dim \alpha = 3, 2, 0$, and, for $\dim \alpha = 1$, $f_{6,18}^1(O, \alpha) = 1$ iff $| \text{st}_3(\alpha; O) | \geq 3$.

d) $f_{6,26}^0(O, \alpha) = 1$ iff $\text{st}_3(\alpha; R^3) \subseteq O$ for any cell $\alpha \in R^3$.

e) $f_{18,6}^0(O, \alpha) = 1$ iff $\alpha \in O$ for $\dim \alpha = 3$, $\alpha \in \text{supp}(O)$ for $\dim \alpha = 1, 2$, and $\text{st}_3(\alpha; O)$ contains either the configuration (3c) or (4e) for $\dim \alpha = 0$.

f) $f_{18,6}^1(O, \alpha) = 1$ iff $\alpha \in O$ for $\dim \alpha = 3$, $\alpha \in \text{supp}(O)$ for $\dim \alpha = 1, 2$, and $| \text{st}_3(\alpha; O) | \geq 3$ and $\alpha \in \text{supp}(O)$ for $\dim \alpha = 0$.

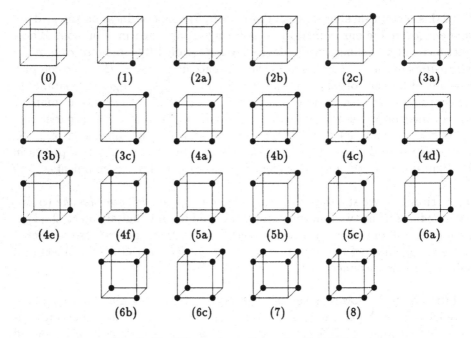

Fig. 2. Possible configurations of a digital object.

g) $f_{18,6}^2(O, \alpha) = 1$ iff $\alpha \in O$ for $\dim \alpha = 3$, $\alpha \in \text{supp}(O)$ for $\dim \alpha = 2$, $\alpha \in \text{supp}(O)$ and $| \text{st}_3(\alpha; O) | = 2, 4$ for $\dim \alpha = 1$, and $\text{st}_3(\alpha; O)$ is one of the configurations (3c), (4b), (4d), (4e), (4f) or $| \text{st}_3(\alpha; O) | \geq 5$ for $\dim \alpha = 0$.

h) $f_{18,18}^0(O, \alpha) = 1$ iff $\alpha \in O$ for $\dim \alpha = 3$, $\alpha \in \text{supp}(O)$ for $\dim \alpha = 1, 2$, and $\text{st}_3(\alpha; O)$ contains the configuration (6c) in Figure 2 for $\dim \alpha = 0$.

i) $f_{18,18}^1(O, \alpha) = 1$ iff $\alpha \in O$ for $\dim \alpha = 3$, $\alpha \in \text{supp}(O)$ for $\dim \alpha = 2$, $\alpha \in \text{supp}(O)$ and $| \text{st}_3(\alpha; O) | = 2, 4$ for $\dim \alpha = 1$, and $\text{st}_3(\alpha; O)$ is one of the configurations (6c), (7) or (8) for $\dim \alpha = 0$.

j) $f_{18,26}^0(O, \alpha) = 1$ iff $\alpha \in O$ for $\dim \alpha = 3$, $\alpha \in \text{supp}(O)$ for $\dim \alpha = 1, 2$, and $| \text{st}_3(\alpha; O) | \geq 7$ for $\dim \alpha = 0$.

k) $f_{26,6}^0(O, \alpha) = 1$ iff $\alpha \in \text{supp}(O)$ for any cell $\alpha \in R^3$; that is, $f_{26,6}^0 = f_{\max}$.

l) $f_{26,18}^0(O, \alpha) = 1$ iff $\alpha \in O$ for $\dim \alpha = 3$, $\alpha \in \text{supp}(O)$ for $\dim \alpha = 2$, $\alpha \in \text{supp}(O)$ and $| \text{st}_3(\alpha; O) | = 2, 4$ for $\dim \alpha = 1$, and $\text{st}_3(\alpha; O)$ is one of the configurations (2c), (4c), (5a), (6a), (6c), (7) or (8) for $\dim \alpha = 0$.

m) $f_{26,26}^0(O, \alpha) = 1$ iff $\alpha \in O$ for $\dim \alpha = 3$, $\alpha \in \text{supp}(O)$ for $\dim \alpha = 2$, $\alpha \in \text{supp}(O)$ and $| \text{st}_3(\alpha; O) | = 2, 4$ for $\dim \alpha = 1$, and $\text{st}_3(\alpha; O)$ is one of the configurations (2c), (6a), (7) or (8) for $\dim \alpha = 0$.

Observe that, in general, the (α, β)-connectedness can be provided by several lighting functions. However, it is not difficult to prove that $f_{6,6}^0$ is the only function providing the (6,6)-connectedness.

It can be readily checked that the simplicial analogue of the whole space $(R^3, f_{\alpha, \beta}^n)$ is the barycentric subdivision of R^3 for all the lighting functions given in Example 2, except for the case $f_{6,6}^0$. Thus, their continuous analogues are always the 3-dimensional Euclidean space $\mathbb{R}^3$. The continuous analogue of the special case $(R^3, f_{6,6}^0)$ is the subset $(\mathbb{Z} \times \mathbb{Z} \times \mathbb{R}) \cup (\mathbb{Z} \times \mathbb{R} \times \mathbb{Z}) \cup (\mathbb{R} \times \mathbb{Z} \times \mathbb{Z}) \subseteq \mathbb{R}^3$.

5 About digital surfaces

In Section 3 we have used our multilevel architecture to show that a merely combinatorial definition, as the notion of O'-connectedness, is a correct counterpart of the topological notion we have at the continuous level. But, in order to define a digital notion, one may proceed along the inverse way. That is, given a topological property P, we can say that a digital object O satisfies the digital counterpart of P by requiring that the continuous analogue $\mathcal{A}_O$ satisfies P. However, doing that, it arises the problem of characterizing property P at a level as close to the logical one as possible.

In this section we will present a case of this method, by defining the notion of digital surface throughout the continuous analogue of objects and then, finding characterizations in the logical level for those digital spaces $(R^3, f_{\alpha,\beta}^n)$ in Example 2.

Definition 12. A digital object S in a digital space (K, f) is said to be a *digital surface* if its continuous analogue $| \mathcal{A}_S |$ is a surface without boundary. We will call S a *f-surface* in case the digital space is (R^3, f).

Kong-Roscoe [6], generalizing the Morgenthaler-Rosenfeld surfaces [11], define in $\mathbb{Z}^3$ the notion of (α, β)-surface for all pairs $\alpha, \beta \in \{6, 18, 26\}$. Next theorem states the characterization in the logical level of the $f_{\alpha,\beta}^n$-surfaces throughout their relation with the corresponding (α, β)-surfaces.

Theorem 13. *The following properties are verified for the lighting functions given in Example 2.*

(1) The family of $f_{\alpha,\beta}^n$-surfaces coincides with the corresponding family of (α, β)-surfaces for the lighting functions: $f_{6,26}^0$, $f_{6,18}^0$, $f_{18,26}^0$, $f_{26,6}^0$, $f_{26,18}^0$ and $f_{26,26}^0$.

(2) The family of $f_{\alpha,\beta}^n$-surfaces is strictly contained in the family of (α, β)-surfaces in the cases: $f_{6,18}^1$, $f_{18,6}^0$, $f_{18,6}^1$ and $f_{18,18}^0$. In addition, the families of $f_{18,6}^0$-surfaces and $f_{18,6}^1$-surfaces coincide.

(3) The families of $(18,6)$-surfaces and $(18,18)$-surfaces are strictly contained in the families of $f_{18,6}^2$-surfaces and $f_{18,18}^1$-surfaces, respectively.

The equality between the families of $f_{26,6}^0$-surfaces and $(26,6)$-surfaces was originally proved in [1]. The same technique can be adapted for each one of the remaining cases in Theorem 13. Notice that we only get an inclusion in (2) because not all the configurations permitted in an (α, β)-surface can appear in an $f_{\alpha,\beta}^n$-surface. Also we only get an inclusion in (3) since an $f_{18,6}^2$-surface may contain the configuration (5b) in Figure 2 which is not permitted in a $(18,6)$-surface.

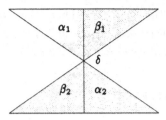

Fig. 3. A polyhedral complex K_0.

Observe that Theorem 13 shows that the family of (α, β)-surfaces, for $(\alpha, \beta) \neq (6,6), (18,6)$, coincides with the family of $f_{\alpha,\beta}^n$-surfaces for some lighting function $f_{\alpha,\beta}^n$. As it was pointed out by Kong-Roscoe in [6], there exists (6,6)-surfaces whose complement is 6-connected. So that, these are not truly digital surfaces. This fact agrees with our point of view because the continuous analogue of the only digital space $(R^3, f_{6,6}^0)$ providing the (6,6)-connectedness is 1-dimensional. Concerning the pair (18,6), it can be proved that there exists no lighting function such that the corresponding family of digital surfaces exactly coincides with the (18,6)-surfaces. These surfaces have, in some sense, an anomalous behaviour with respect the others. In a future work we will intend a more detailed analysis of them; and moreover we also plan to find lighting functions gathering other definitions of digital surface, as those due to Malgouyres [9,10].

6   Final discussions

In this last section we want to discuss some interesting points of our framework as it is compared with other approaches to digital topology.

Firstly, we focus our attention on our notion of connectedness (Definition 5), which is slightly different from that normally used in abstract cell complexes. To explain this difference, let us consider the polyhedral complex K_0 in Fig. 3, where we will distinguish two digital objects $A = \{\alpha_1, \alpha_2\}$ and $B = \{\beta_1, \beta_2\}$, and the vertex δ. Let us also consider the lighting function f ("membership rule" in Kovalevsky's terminology) defined on K_0 by $f(O, \gamma) = 1$ for any digital object $O \subseteq \text{cell}_n(K_0)$ and any cell $\gamma \in \text{supp}(O)$ with $\gamma \neq \delta$, and $f(O, \gamma) = 0$ otherwise. Observe that the "membership rule" f associates the 0-cell δ only to complements of objects; see the comment after Definition 1.

Let now consider the digital space (K_0, f); i.e., the set $\text{cell}_n(K_0)$ of all pixels in K_0 together with the set of lower dimensional cells associated to it

According to the arcwise connectedness usually defined on cell complexes (see Definition 4 in [8]), the digital space (K_0, f) has two connected components whose sets of pixels are $\{\alpha_1, \beta_1\}$ and $\{\alpha_2, \beta_2\}$ respectively. Moreover, if we consider B as an object, its complement A is connected although A meets both components $\{\alpha_1, \beta_1\}$ and $\{\alpha_2, \beta_2\}$. This situation is avoided with our notion of connectedness.

The crucial point is that, in the usual definition, each lower dimensional cell γ associated to the complement $\text{cell}_n(K) - O$ of an object O is connecting the pixels

(n-cells) in the star of γ in $\mathrm{cell}_n(K) - O$. In contrast our Definition 5 expresses only that γ is not a cut-point for $\mathrm{st}_n(\gamma; \mathrm{cell}_n(K) - O)$. In order to ensure that γ connects $\mathrm{st}_n(\gamma; \mathrm{cell}_n(K) - O)$ it is required in addition that γ is lighted in the global object $\mathrm{cell}_n(K)$; i.e., γ connects the pixels in $\mathrm{st}_n(\gamma; \mathrm{cell}_n(K))$.

In relation to the notion of connectedness used in the graph-based approach to digital topology, it may seem puzzling the existence of lighting functions providing the (α, β)-connectedness, for $\alpha, \beta \in \{18, 26\}$. These pairs are usually discarded on the ground that their restrictions on the grid $\mathbb{Z}^2 \times \{0\} \subseteq \mathbb{Z}^3$ produce the paradoxical $(8, 8)$-connectedness. However, this is not the case of our lighting functions $f_{\alpha, \beta}$.

For instance, consider the lighting function $f_{18,18}^0$ given in Example 2(h) and the subcomplex $R_0^3 = \{\alpha \in R^3 : \alpha \le \sigma, c(\sigma) \in \mathbb{Z}^2 \times \{0\}\}$ of R^3. It is easy to show that the restriction of $f_{18,18}^0$ to the plane R_0^3, given by $f_{18,18}^0 \mid_{R_0^3} (O, \alpha) = 1$ if and only if $\alpha \in \mathrm{supp}(O)$ for any object $O \subseteq \mathrm{cell}_3(R_0^3)$ and any cell $\alpha \in R_0^3$, is a lighting function providing the $(8, 4)$-connectedness.

Next, we will illustrate how is this possible through an example. Let consider the digital object $O = \{\sigma_n \in \mathrm{cell}_3(R^3) : c(\sigma_n) = (n, n, 0), n \in \mathbb{Z}\}$ in the digital space $(R^3, f_{18,18}^0)$ consisting of a diagonal line in the digital plane R_0^3, and let $\tau_1, \tau_2 \in \mathrm{cell}_3(R^3) - O$ be the two only 3-cells sharing the edge $\gamma = \sigma_1 \cap \sigma_2$. According to the $(18, 18)$-adjacency both pairs of 3-cells, σ_1, σ_2 and τ_1, τ_2, should be adjacent. Indeed, σ_1 and σ_2 are adjacent (i.e., $\emptyset$-adjacent) in O through their common edge γ, while τ_1 and τ_2 are O-adjacent in $\mathrm{cell}_3(R^3) - O$ through any one of the extremes of γ. On one hand, O and $\mathrm{cell}_3(R^3) - O$ are connected and O-connected objects respectively, as well as they are 18-connected. On the other hand, consider the same object O and cells τ_1, τ_2 in the digital subspace $(R_0^3, f_{18,18}^0 \mid_{R_0^3})$. In this space, σ_1 and σ_2 are again adjacent in O through γ. But now, τ_1 and τ_2 are not O-adjacent in $\mathrm{cell}_3(R_0^3) - O$ because the extremes of γ do not belong to $\mathrm{supp}(R_0^3)$. Thus, O is connected as an object in the digital subspace $(R_0^3, f_{18,18}^0 \mid_{R_0^3})$, but its complement $\mathrm{cell}_3(R_0^3) - O$ is not O-connected.

Finally, we are going to justify briefly why a multilevel architecture seems to us very suitable for the development of digital topology.

The goal of digital topology is to analize and to study topological properties on digital objects, under the assumption that, although these objects have strictly a discrete nature, they are perceived as continuous objects. Because of this, any framework for digital topology should contain at least these two levels: a digital one, in which digital images can be easily processed, and a continuous level, where digital methods and results can be justified in accordance with the continuous perception of objects. Obviously these levels are of very different nature and we have considered convenient to introduce some other levels which, in conjunction with suitable transformations, make easier the translation of notions and results between the digital and continuous levels.

This architecture has allowed us to define, in a very natural way, a continuous analogue close enough to the perception of objects. And, what has greater importance, it makes possible to reuse knowledges and experiences of continuous topology, by translating them to the digital level throughout the whole architecture. In addition, it is worth pointing out that our architecture gathers in its levels the somehow scattered proposals by other authors. Actually, the device level corresponds to Kovalevsky's approach, the logical level falls within the graph-based models due to Rosenfeld and others, and finally, Khalimsky's spaces are a particular case of the conceptual level.

Acknowledgements:
This work was partially supported by the project DGICYT PB96-1374.

References

1. R. Ayala, E. Domínguez, A.R. Francés, A.Quintero, J. Rubio. On surfaces in digital topology. *Proc. of the 5th Workshop on Discrete Geometry for Computer Imagery DGCI'95.* (1995) 271-276.
2. R. Ayala, E. Domínguez, A.R. Francés, A.Quintero. Determining the components of the complement of a digital $(n-1)$-manifold in $\mathbb{Z}^n$. *Proc. of the 6th Int. Workshop on Discrete Geometry for Computer Imagery DGCI'96. Lectures Notes in Computer Science.* 1176(1996) 163-176.
3. E. Domínguez, A.R. Francés, A. Márquez. A Framework for Digital Topology. *Proc. of the IEEE Int. Conf. on Systems, Man, and Cybernetics.* 2(1993) 65-70.
4. J. Françon. On recent trends in discrete geometry in computer science. *Proc. of the 6th Int. Workshop on Discrete Geometry for Computer Imagery DGCI'96. Lectures Notes in Computer Science.* 1176(1996) 163-176.
5. E. Khalimsky, R. Kopperman, P.R. Meyer. Computer Graphics and connected topologies on finite ordered sets. *Topology and its Applications.* 36(1990) 1-17.
6. T.Y. Kong, A.W. Roscoe. Continuous Analogs of Axiomatized Digital Surfaces. *Computer Vision, Graphics, and Image Processing.* 29(1985) 60-86.
7. T.Y. Kong, A. Rosenfeld. Digital Topology: Introduction and Survey. *Computer Vision, Graphics, and Image Processing.* 48(1989) 357-393.
8. V.A. Kovalevsky. Finite topology as applied to image analysis. *Computer Vision, Graphics, and Image Processing.* 46(1989) 141-161.
9. R. Malgouyres. A definition of surfaces of $\mathbb{Z}^3$. *Proc. of the 3th Workshop on Discrete Geometry for Computer Imagery DGCI'93.* (1993) 23-34.
10. G. Bertrand, R. Malgouyres. Some topological properties of Discrete Surfaces. *Proc. of the 6th Int. Workshop on Discrete Geometry for Computer Imagery DGCI'96. Lectures Notes in Computer Science.* 1176(1996) 325-336.
11. D.G. Morgenthaler, A. Rosenfeld. Surfaces in Three-Dimensional Digital Images. *Information and Control.* 51(1981) 227-247.
12. C.P. Rourke, B.J. Sanderson. *Introduction to Piecewise linear topology.* Ergebnisse der Math., 69. Springer, 1972.

Digital Topologies Revisited:
An Approach Based on the Topological
Point-Neighbourhood*

Pavel Ptak[1], Helmut Kofler[2], Walter Kropatsch[2]

[1]Center for Machine Perception
CMP – Czech Technical University
Faculty of Electrical Engineering
Karlovo nam. 13
121 35 Praha 2, Czech Republic
ptak@math.feld.cvut.cz

[2]Pattern Recognition and
Image Proc. Group – PRIP
Vienna University of Technology
Treitlstr. 3,
A-1040 Vienna, Austria
{krw, kof}@prip.tuwien.ac.at

Abstract. Adopting the point-neighbourhood definition of topology, which we think may in some cases help acquire a very good insight of digital topologies, we unify the proof technique of the results on 4-connectedness and on 8-connectedness in $\mathbf{Z}^2$. We also show that there is no topology compatible with 6-connectedness. We shortly comment on potential further use of this approach.

Keywords: Image processing, digital topology, adjacency, path-connectedness and topological connectedness in $\mathbf{Z}^2$.

1 Introduction and Basic Definitions

The domain of a *digital picture* can be viewed as a subset of $\mathbf{Z}^2$, where $\mathbf{Z}$ stands for the set of integers, together with some adjacency neighbourhood structure [2, 7, 6] assigned to each point. Thus, for instance, we may talk on 8-adjacency neighbourhood structure (in shorthand, 8-structure) if each point $(x, y) \in \mathbf{Z}^2$ is given the adjacency neighbourhood

$$(x-1, y-1) \ (x, y-1) \ (x+1, y+1)$$
$$(x, y-1) \quad (x, y) \quad (x, y+1)$$
$$(x-1, y+1) \ (x+1, y) \ (x+1, y+1)$$

In the sequel, let us assume as plausible that *the adjacency neighbourhood structure is homogeneous* (i.e., for each (x_0, y_0), the natural translation $(x, y) \rightarrow (x + x_0, y + y_0)$ is an adjacency isomorphism) and *symmetric* with respect to the point (x, y).

---

* This work was supported by a grant from the Austrian National Fonds zur Förderung der wissenschaftlichen Forschung (No. S7002MAT).

We would like to demonstrate that the point-neighbourhood definition of topology adopted here provides us with a good method for deciding when (if) there is a topology compatible with adjacency. We will present different proofs of previously known results (see [2, 6, 7, 9]) and slightly extend them.

We also assume that the adjacency neighbourhood never exceeds the 8-neighbourhood, but note the result holds also for any larger neighbourhood.

In the following we give some elementary definitions. In *section 2* we present the different *adjacency neighbourhood structures*. *Section 3* contains the main results. Finally some conclusions are given in *section 4*.

1.1 Path-Connectedness in S

Suppose that $\mathbf{Z}^2$ is given a (homogeneous and symmetric) neighbourhood structure. In order to refer to this structure, let us call it S. Suppose that p, q are points of $\mathbf{Z}^2$. By an *S-path from p to q* we understand a finite sequence of points $p = p_1, p_2, \ldots, p_n = q$ such that p_i is a neighbour of p_{i-1} $(1 < i \leq n)$ in the structure S. Let us call the points p, q *S-related* if there is an S-path from p to q.

Let X be a subset of $\mathbf{Z}^2$. Since the relation of being S-related, for a given S, is obviously an equivalence relation on X. This relation gives rise to a partition of X into classes of S-related elements. Let us call each class of this equivalence an *S-component*.

A natural question arises ([8, 2], etc.) if (when) the S-components can be obtained as the components of a connectedness relation of a topology. Let us view basic notions we need for pursuing this question. Out of several possible definitions of (classical) topology, the definition involving the point-neighbourhood structure may best serve the purpose.

1.2 Topological Connectedness

Definition (Topological space): *Let P be a set. Let us assign to each $x \in P$ a set $\mathcal{U}(x)$ of subsets of P which is subject to the following conditions:*

(i) *if $U \in \mathcal{U}(x)$, then $x \in U$,*

(ii) *if $U \in \mathcal{U}(x)$ and $U \subset V$, then $V \in \mathcal{U}(x)$,*

(iii) *if $U, V \in \mathcal{U}(x)$, then $U \cap V \in \mathcal{U}(x)$,*

(iv) *if $U \in \mathcal{U}(x)$, then there exists $V \in \mathcal{U}(x)$ such that, for each $y \in V$, $U \in \mathcal{U}(y)$.*

The set P together with the assignment $\mathcal{U}(x)$, for each $x \in P$, is called a topological space. *The sets $U \in \mathcal{U}(x)$ are called* topological neighbourhoods *of x. We denote the assignment $x \to \mathcal{U}(x)$ by t - the* topology. *So we can refer to the couple (P, t), meaning the corresponding topological space.*

It should be noted that this definition of topological space is equivalent to the "base–for–open–sets" definition or to the "closure" definition. This can be easily verified by a straightforward translation (see e. g. [1]). One should also observe that it is the axiom (iv) which is usually responsible for inconveniences when one looks for "suitable" topologies (this can be compensated by a possibility to apply the topological result back into the real model).

Let (P, t) be a topological space defined in the sense of previous definition (i. e. via point–neighbourhoods) and let X be a subset of P. We say that (X, t_1) is a *topological subspace* of (P, t) if, for each $x \in X$, the set $\mathcal{U}(x) \cap X$ is the set of all neighbourhoods of x in the topology t_1. It can be seen easily that (X, t_1) is then a topological space in its own right. Moreover, if (P, t) is a topological space and Y, V are subsets of P with $Y \subset V$, then if $(Y, t_1), (V, t_2)$ are topological subspaces of (P, t), then (Y, t_1) is a topological subspace of (V, t_2).

Definition (Topological connectedness): *Let (P, t) be a topological space. We say that (P, t) is disconnected if there are two disjoint sets R, S such that $R \cup S = P$ and, moreover, for each $r \in R$ and each $s \in S$ the set R is a neighbourhood of r and S is a neighbourhood of s. The space (P, t) is said to be connected if it is not disconnected. Finally, a subset X of (P, t) is said to be connected if the subspace (X, t_1) of (P, t) is connected.*

Obviously, a connected topological space may have plenty of disconnected subspaces. For instance, the set $(-\infty, 0) \cup (0, +\infty)$ or the set $\{1, 2, 3, \ldots\}$ are obviously disconnected subspaces of the space R of reals (resp. the topology induced by one of the equivalent n-norms $n = 1, 2, \infty$). Also, the set Q of all rational numbers is also disconnected – we can write $Q = Q_1 \cup Q_2$, where $Q_1 = \{q \in Q \mid q < \sqrt{2}\}$ and $Q_2 = \{q \in Q \mid q > \sqrt{2}\}$. The last example shows that the question on deciding about connectedness of a subspace may be sometimes nontrivial.

2 Neighbourhood Structures on $\mathbf{Z}^2$

Definition (Compatibility): *Let S denote a given adjacency structure on $\mathbf{Z}^2$ and let t be a topology on $\mathbf{Z}^2$. We say that t is compatible with S if for each $X, X \subset \mathbf{Z}^2$, the set X is connected (with respect to t) if and only if X is S-connected.*

Let us employ the following definition (see also [3]). Let (P, t) be a topological space. We say that (P, t) is *locally finite* if each point $x \in P$ possesses the finite neighbourhood $U \in \mathcal{U}(x)$, i.e. $|U| < \infty$. Obviously, if (P, t) is locally finite then *each point $x \in P$ possesses a smallest neighbourhood* in t and this neighbourhood is finite. This follows from the closedness of the neighbourhood under the formation of intersections. An important fact in our considerations is this: If $U(x)$ is the smallest (topological) neighbourhood of x, then $U(x)$ must also be a neighbourhood of all points in $U(x)$. This immediately follows from the point-neighbourhood definition of topology (the axiom (iv)).

We will now investigate all possible (homogeneous and symmetric) adjacency neighbourhood structures in $\mathbf{Z}^2$ (see the figure below) and formulate results on the compatible topologies. We will in turn take up the 0-adjacency, 2-adjacency, 4-adjacency, 6-adjacency and 8-adjacency.

(i)   •   0-adjacency (to each point $(x, y) \in \mathbf{Z}^2$ the only point (x, y) is adjacent)

(ii)   •——•——•   2-adjacency (to each point $(x, y) \in \mathbf{Z}^2$ the points $(x - 1, y)$ and $(x + 1, y)$ are adjacent)

(iii)   4-adjacency (to each point $(x, y) \in \mathbf{Z}^2$ the points $(x - 1, y), (x + 1, y), (x, y - 1), (x, y + 1)$ are adjacent)

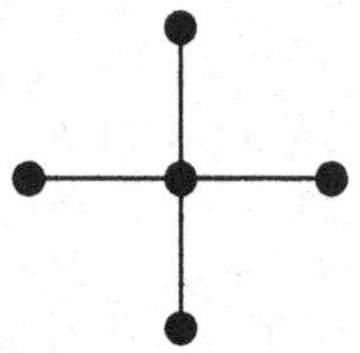

(iv)   6-adjacency (to each point $(x, y) \in \mathbf{Z}^2$ the points $(x - 1, y), (x + 1, y), (x, y - 1), (x, y + 1),$ $(x - 1, y - 1), (x + 1, y + 1)$ are adjacent)

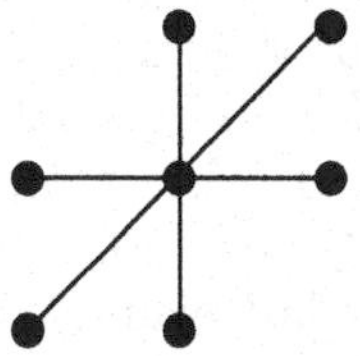

(v)   8-adjacency (to each point $(x, y) \in \mathbf{Z}^2$ the points $(x - 1, y), (x + 1, y), (x, y - 1), (x, y + 1),$ $(x - 1, y - 1), (x + 1, y + 1), (x - 1, y + 1),$ $(x + 1, y - 1)$ are adjacent).

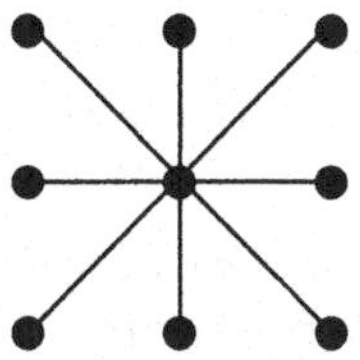

Remark: The n-adjacencies, n is odd, are not homogeneous and 2-point connections (1-adjacency) can be compared only with a trivial topology like the discrete topology.

3   Compatible Topologies on $\mathbf{Z}^2$

Theorem 1: *There is exactly one topology which is compatible with the 0-adjacency. This topology is the discrete topology on $\mathbf{Z}^2$.*

Proof: The discrete topology making each point $(x,y) \in \mathbb{Z}^2$ a neighbourhood of (x,y) is clearly compatible with 0-adjacency.

Theorem 2: *There are infinitely many topologies which are compatible with the 2-adjacency in $\mathbb{Z}^2$. Among those there are infinitely many, which are locally finite.*

Proof: For a fixed $x \in \mathbb{Z}$, we obviously have a topology, t_x, on the subspace $T_x = \{(x,y) \mid y \in \mathbb{Z}\}$ which is compatible with the 2-adjacency on the set T_x. Indeed, it suffices to take $\mathcal{U}(x) = \{U \subset T_x \mid \{(x,y)\} \subset U\}$ provided y is odd, $\mathcal{U}(x) = \{U \subset T_x \mid \{(x,y-1),(x,y),(x,y+1)\} \subset U\}$ provided y is even. Note that the above neighbourhoods contain two types as smallest and finite neighbourhood (*i.e. the 1-dimensional* Marcus-Wyse *topology*).

Since the roles of odd and even numbers are obviously interchangeable and since the adjacency connectedness (2-connectedness) of "different levels" do not affect each other, we can take, for each $x \in \mathbb{Z}$, one of the two topologies on T_x, obtaining a topology that is compatible with 2-adjacency. Since we have infinitely many combinations at our disposal, the result is proved.

Remark (not locally finite topologies): A question of separate purely topological curiosity may arise whether we can construct, for the 2-adjacency, a topology which is compatible with the adjacency and which is not locally finite. This seems to be possible – the standard ultrafilter construction can be applied in this case or more easily the Frechet-filter of the infinite intervals $\{x \in \mathbb{Z} \mid x \geq z\}$, $z \in \mathbb{Z}$. These topologies are, however, hardly relevant to digital pictures.

Let us now consider the 4-adjacency. The following result, which is due to [9] and [6] is in force. We will show how one obtains this result in the point–neighbourhood formalism. The little auxiliary results stated as Observations 3.1, 3.2 may be of certain value in their own right.

Theorem 3: *There are 2 topologies which are compatible with the 4-adjacency — the 2-dimensional* Marcus-Wyse *topologies τ [9]:*

$$U \in \tau \quad \equiv \quad \begin{cases} U(x,y) & : \text{ if } x+y \text{ is even} \quad (\text{resp. odd}), \\ \{(x,y)\} & : \text{ else,} \end{cases}$$

with $U(x,y) = \{(x,y),(x,y-1),(x,y+1),(x+1,y),(x-1,y)\}$.

Proof: The result easily follows from the following two observations.

Observation 3.1: Each topology which is compatible with 4-adjacency is locally finite. Moreover, if $U(x,y)$ is the smallest neighbourhood of (x,y) in a topology compatible with 4-adjacency, then

$$U(x,y) \quad \subseteq \quad \{(x,y),(x,y-1),(x,y+1),(x+1,y),(x-1,y)\}.$$

Proof: Let $(x,y) \in \mathbb{Z}^2$ and let us show that the point (x,y) must have a finite neighbourhood.

Let $X_1 = \{(x,y)\}$ and $X_2 = \{(u,v) \in \mathbf{Z}^2 | \; |x - u| + |y - v| \geq 2\}$, see Fig. 1.

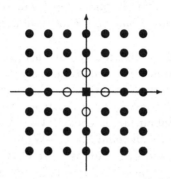

Fig. 1. The two sets $X_1 = \{\blacksquare\}$ and $X_2 = \{\bullet\}$

Then both X_1 and X_2 are 4-connected but $X_1 \cup X_2$ is not. It follows that there is a topological neighbourhood of (x, y) in any topology compatible with 4-adjacency which is disjoint with X_2. In other words, there is a topological neighbourhood, $U(x, y)$, such that $U(x, y) \subseteq \mathbf{Z}^2 - X_2$. This is what we wanted to show.

Observation 3.2: Let $(x, y) \in \mathbf{Z}^2$ and let t be a topology compatible with 4-adjacency. Let $U(x, y)$ be the smallest neighbourhood of (x, y) in t. Then either $U(x, y) = \{(x, y)\}$ or $U(x, y) = \{(x, y), (x, y - 1), (x, y + 1), (x + 1, y), (x - 1, y)\}$.

Proof: By the previous observation, $U(x, y) \subseteq \{(x, y), (x, y - 1), (x, y + 1), (x + 1, y), (x - 1, y)\}$. Suppose that $U(x, y) \neq \{(x, y)\}$. Then, without a loss of generality, $(x - 1, y) \in U(x, y)$. Suppose now that $(x, y + 1) \notin U(x, y)$ (again, one argues analogously if there is another edge than $(x, y+1)$ outside of $U(x, y)$). Since the set $\{(x-1, y), (x, y+1)\}$ is not 4-connected and the set $\{(x, y), (x, y+1)\}$ is 4-connected, we infer that the smallest neighbourhood of $(x, y + 1)$, some set $U(x, y + 1)$, must not contain the point $(x - 1, y)$ and must contain the point (x, y). Consequently considering other edges analogously,

$$U(x, y) \cap U(x, y + 1) = \{(x, y)\}.$$

But since $U(x, y+1)$ is a topological neighbourhood, it must be also a topological neighbourhood of the point (x, y). It follows that $U(x, y) \cap U(x, y + 1)$ must be a topological neighbourhood of the point (x, y).
But $U(x, y) \cap U(x, y+1) = \{(x, y)\}$ which is a contradiction. This completes the proof of Observation 3.2.

Let us return to the proof of Theorem 3. Let t be a topology which is compatible with the 4-adjacency. It is obvious that the singleton sets $\{(x, y)\}$ cannot constitute the neighbourhoods for all $(x, y) \in \mathbf{Z}^2$ (we would obtain the discrete topology; the discrete topology is obviously not compatible with 4-adjacency). It is also obvious that the sets

$$\{(x, y), (x - 1, y), (x + 1, y), (x, y - 1), (x, y + 1)\}$$

cannot be the smallest neighbourhoods for all $(x, y) \in \mathbf{Z}^2$ (we would not have a topology at all). Thus, in every topology compatible with 4-adjacency we must have, for some points, both the singleton smallest neighbourhoods and the "star–like" neighbourhoods. Having found the necessary conditions for a compatible topology, the rest consists of an easy inductive argument already presented in [9].

Choose, for instance, the point $(0, 0)$. Then either the set $U_1(0, 0) = \{(0, 0), (-1, 0), (1, 0), (0, 1), (0, -1)\}$ or the set $U_2(0, 0) = \{(0, 0)\}$ must be the smallest neighbourhood of $(0, 0)$.

In the former case, the smallest neighbourhoods of the points $(-1, 0), (1, 0), (0, -1), (0, 1)$ must necessarily be the singleton sets, the smallest neighbourhoods of the points $(-1, -1), (-1, 1), (1, -1), (1, 1)$ must be the "star" sets, and so on. In the latter case, the smallest neighbourhoods of the points $(-1, 0), (1, 0), (0, -1), (0, 1)$ must be the "star" sets, the smallest neighbourhoods of the points $(-1, -1), (-1, 1), (1, -1), (1, 1)$ must be the singleton sets, and so on.

Consequently, there are only two topologies on $\mathbf{Z}^2$ which are compatible with the 4-adjacency – either the Marcus-Wyse topology or the topology obtained from it by the shift $(x, y) \to (x, y + 1)$. The Marcus-Wyse topology allows for a simple description as also the present consideration shows: the smallest neighbourhood of (x, y) is a singleton set provided $x + y$ is even, and the smallest neighbourhood of (x, y) is a star set provided $x + y$ is odd.

The rest would consist in checking that, indeed, the Marcus-Wyse topology is compatible with 4-adjacency. This is not difficult and has been done in detail in [9]. The proof is complete.

Let us now consider the 6-adjacency (see the schema in the figure (iv)). In this case the search for compatible topology would be in vain.

Theorem 4: *There is no topology on $\mathbf{Z}^2$ which is compatible with the 6-adjacency.*

Proof: Suppose that t is a topology compatible with the 6-adjacency. By the very same reasoning we employed in Observations 3.1, 3.2 we can derive the following results:

(i) The topology t is locally finite,

(ii) If $(x, y) \in \mathbf{Z}^2$, then the smallest neighbourhood of (x, y) in t is either the singleton set $\{(x, y)\}$ or the whole 6-star set

$$\{(x, y), (x - 1, y), (x + 1, y), (x, y - 1), (x, y + 1), (x - 1, y - 1), (x + 1, y + 1)\}.$$

There must be a point in $\mathbf{Z}^2$ with the proper 6-adjacency neighbourhood. Let us denote it again by (x, y). It follows that the points $(x + 1, y)$ and $(x + 1, y + 1)$ must have the singleton set neighbourhoods. This is absurd since the set $\{(x+1, y), (x+1, y+1)\}$ is 6-connected. The proof is complete.

The following corollary to the previous result can be viewed as another proof of the result by Chassery [2] and L. Latecki [7].

Theorem 5: *There is no topology on $\mathbf{Z}^2$ which is compatible with the 8-adjacency.*
Proof: It is easily seen that if S_1 and S_2 are two adjacency relations on $\mathbf{Z}^2$ and S_2 is finer then S_1, then the absence of a locally finite compatible topology for S_1 implies the absence of a locally finite compatible topology for S_2. Since there is no topology compatible with 6-adjacency, there is no topology compatible with the 8-adjacency.

4 Conclusion

We have completed the tour over all possible "nice" adjacencies in $\mathbf{Z}^2$. Presumably the next step is testing the suitability of the point–neighbourhood approach is the examination of concrete (finite) configurations of points in $\mathbf{Z}^2$ and, of course, the digital topologies in $\mathbf{Z}^n$ for $n \geq 3$. We intend to pursue this elsewhere. It should be observed, however, that it does not seem possible to analyze $\mathbf{Z}^n$ with the help of viewing $\mathbf{Z}^n$ as a Cartesian product of $\mathbf{Z}^m$ for m smaller than n. This can be graphically seen even for $\mathbf{Z}^2$. Indeed, none of the adjacencies on $\mathbf{Z}^2$ with the exception of the discrete one is obtained as a "Cartesian product" of adjacencies on $\mathbf{Z}$. We may however obtain nontrivial (non-homogeneous) adjacencies this way or, in other words, homogeneity may be a too restricting property for digital topologies. In particular if the data stem from a projection of a higher dimensional space (e.g. 3-dim) onto a lower dimension (e.g. 2-dim). In the case of digital images, we may seek the topological properties of the 3-dim objects in the 2-dim image rather than establishing adjacency across object boundaries. Such occluding boundaries represent discontinuities of the image function and adjacent pixels may correspond to 3-dim points of different objects which are far apart in reality.
If, for instance, we take the 2-adjacency on $\mathbf{Z}$ and multiply it with each other, we obtain an adjacency on $\mathbf{Z}^2$ which is the 4-adjacency on the points of the type $(even, even)$, the 2-adjacency on the points of the type $(even, odd)$ and $(odd, even)$ − in the former case vertically and in the latter case horizontally, and the 1-adjacency on the points of the type (odd, odd). Since the 2-adjacency on $\mathbf{Z}$ allows for a compatible topology, so does our "mixed" adjacency on $\mathbf{Z}^2$, which directly leads to the abstract cellular complexes of Kovalevsky [3, 4].
This survey is part of an ongoing research project with two primary goals to extend the results to *higher dimensions* and to *irregular adjacency neighbourhood structures* [5].

References

1. E. Cech: *Topological Spaces*, Interscience, Wiley, New York, 1966.
2. J. M. Chassery: *Connectivity and consecutivity in digital pictures*, Computer Graphics and Image Processing 9, 294–300, 1979.
3. E. Khalimsky, R. Kopperman and P. R. Meyer: *Computer graphics and connected topologies on finite ordered sets*, Topology Appl. 36, 117, 1980.
4. V. A. Kovalevsky: *Finite topology as applied to image analysis*, Computer Vision, Graphics and Image Processing 46, 141-161, 1989.
5. W. G. Kropatsch: *Equivalent Contraction Kernels and the Domain of Dual Irregular Pyramids*, Technical Report PRIP-TR-042, TU-Vienna, 1996.
6. L. Latecki: *Digitale und Allgemeine Topologie in der bildhaften Wissensrepräsentation*, Ph.D.-Thesis, Hamburg, 1992.
7. L. Latecki: *Topological connectedness and 8-connectedness in digital pictures*, Computer Vision, Graphics and Image Processing: Image Understanding 57, 261–262, 1993.
8. A. Rosenfeld: *Digital topology*, Am. Math. Monthly 86, 621–630, 1979.
9. F. Wyse and D. Marcus et al.: *Solution to Problem 5712*, Am. Math. Monthly 77, 1119, 1970.

Features

The Euler Characteristic of Discrete Object

Atsushi Imiya* and Ulrich Eckhardt

Dept. of Applied Mathematics, University of Hamburg
Bundessstrasse 55, 20146 Hamburg, Germany

Abstract. We introduce the curvature indexes of the boundary of a discrete object, and using these indexes of points, we define vertex angles of discrete surfaces as an extension of the chain codes for digital curves. Next, we prove a relation between the number of points on the surface and the genus of a discrete object. This is the angular Euler characteristic of a discrete object. These relations derive a parallel algorithm for the computation of the Euler characteristic of a discrete object.

1 Introduction

In this paper we introduce the curvature indexes of points on the boundary of a discrete object using the neighborhood decomposition and the curvature indexes of planar digital curves. The decomposition of the three-dimensional neighborhood to a collection of two-dimensional neighborhoods reduces the combinatorial properties of a discrete object to the collections of combinatorial properties of planar patterns [1]. Using these indexes of points, we define the vertex angles of a discrete surface as an extension of the chain code of digital curves [1]. These indexes provide a relation between the point configurations on the boundary and the Euler characteristic of a discrete object. This is the angular Euler characteristic [5] for a 6-connected discrete object. This relation automatically derives a parallel algorithm for the computation of the Euler characteristics of a discrete object. In this paper a hole is a point set which is encircled by an object. A tunnel is a point set which connects point sets which are locally separated by an object. Denoting a wall which transform a tunnel to a pair of wells of a discrete object, we derive an equation for the computation of the Euler characteristic of a discrete object using points on the boundary [4].

Recently in computer vision, shape reconstruction from multiview images is concerned from viewpoints of the mathematical theory such as the multilinear form for corresponding points of a series of images, and practical applications such as the construction of geometric date of three-dimensional object from a series of tow-dimensional images. This relation among images provides methods for the estimation and generation of new images from observed data. The reconstruction of shape from multiview images is considered as an interpolation problem

---

* Permanent Address: Dept. of Information and Computer Sciences, Chiba University, 1-33 Yayoi-cho, Inage-ku, Chiba 263, Japan, imiya@ics.tj.chiba-u.ac.jp. While staying in Germany, the first author was supported by the Telecommunications Advancement Foundation.

in three-dimensional Euclidean space from a collection of two-dimensional data on imaging planes. Since each image contains noises, errors in the reconstructed data are the collection of errors of projected images. This means the signal-to-noise ratio in a reconstructed object is usually worse than the signal-to-noise ratio of each image. These instabilities cause inaccuracies for the configuration of reconstructed points. For the reduction of these geometric instabilities of re-constructed spatial points, it is necessary to consider the topology of an object, since topological properties are expressed by Boolean values which is stable for the calculation in digital computers [6]. In digital computer, an object is ex-pressed as a discrete object which is a collection of lattice points. Therefore, it is desired to define topological characteristics of discrete objects.

The Euler characteristic is a combinatorial relation among vertices, edges, faces, and tunnels of a polyhedron. The Euler characteristic plays important role in *differential geometry in the large* [7], and digital image processing. The numbers of holes and connected components of a planar discrete binary image derive the Euler characteristic of an image. For topological preserving trans-formations such as deformation and skeltonization, it is desired to check the topology characteristics of an image using the Euler characteristic [3] [8]. For digital binary images a relation among the chin codes, boundary elements, and the Euler characteristic is proven [9]. Furthermore, an algorithm for the com-putation of the Euler characteristic based on this property was developed. Bieri and Nef [15] proposed an algorithm for the computation of the Euler character-istic of a n-dimensional binary object as an application of the cell decomposition of a polytope. In reference [10], Toriwaki and coworkers applied this idea for the definition of the Euler characteristic of a discrete object. Furthermore, they derived a Boolean function for the computation of the Euler characteristic [11]. These results are the discrete versions of usual expression of Euler characteristic by the cell decomposition. Lee, Poston, and Rosenfeld [12] proposed a method for the definition of holes and tunnels of discrete objects using the theory of *cuts*, that is, they dealt with a discrete version of a topological property between the minimum number of cuts which eliminate holes and tunnels and the number of holes and tunnels of an object. The numbers of holes and tunnels are called the genuses of a planar pattern and a spatial object, respectively. Points on the boundary of an object also derive the topological characteristics of object.

2   Connectivity and Neighborhood

Setting $\mathbf{R}^2$ and $\mathbf{R}^3$ to be two- and three- dimensional Euclidean spaces, we express vectors in $\mathbf{R}^2$ and $\mathbf{R}^3$ as $x = (x, y)^\top$ and $x = (x, y, z)^\top$, where $\top$ is the transpose of vectors. Setting $\mathbf{Z}$ to be the set of all integers, the two- and three-dimensional discrete spaces $\mathbf{Z}^2$ and $\mathbf{Z}^3$ are sets of points such that both x and y are integers and all x, y, and z are integers, respectively.

On $\mathbf{Z}^2$ and in $\mathbf{Z}^3$

$$\mathbf{N}^4((m, n)^\top) = \{(m \pm 1, n)^\top, (m, n \pm 1)^\top\} \tag{1}$$

and

$$\mathbf{N}^6((k,m,n)^\top) = \{(k \pm 1, m, n)^\top, (k, m \pm 1, n)^\top, (k, m, n \pm 1)^\top\} \qquad (2)$$

are the planar 4-neighborhood of point $(m,n)^\top$ and the spatial 6-neighborhood of point $(k,m,n)^\top$, respectively. In this paper, we assume the 4-connectivity on $\mathbf{Z}^2$ and the 6-connectivity in $\mathbf{Z}^3$.

Setting one of x, y, and z to be a fixed integer, we obtain two dimensional sets of lattice points such that

$$\mathbf{Z}_1^2((k,m,n)^\top) = \{(k,m,n)^\top | \exists k, \forall m, \forall n \in \mathbf{Z}\}, \qquad (3)$$

$$\mathbf{Z}_2^2((k,m,n)^\top) = \{(k,m,n)^\top | \forall k, \exists m, \forall n \in \mathbf{Z}\}, \qquad (4)$$

and

$$\mathbf{Z}_3^2((k,m,n)^\top) = \{(k,m,n)^\top | \forall k, \forall m, \exists n \in \mathbf{Z}\}. \qquad (5)$$

These two dimensional discrete spaces are mutually orthogonal. Denoting

$$\mathbf{N}_1^4((k,m,n)^\top) = \{(k, m \pm 1, n)^\top, (k, m, n \pm 1)^\top\}, \qquad (6)$$

$$\mathbf{N}_2^4((k,m,n)^\top) = \{(k \pm 1, m, n)^\top, (k, m, n \pm 1)^\top\}, \qquad (7)$$

and

$$\mathbf{N}_3^4((k,m,n)^\top) = \{(k \pm 1, m, n)^\top, (k, m \pm 1, n)^\top\}, \qquad (8)$$

the relationship

$$\mathbf{N}^6((k,m,n)^\top) = \mathbf{N}_1^4((k,m,n)^\top) \cup \mathbf{N}_2^4((k,m,n)^\top) \cup \mathbf{N}_3^4((k,m,n)^\top), \qquad (9)$$

holds, since $\mathbf{N}_i^4((k,m,n)^\top)$ is the 4-neighborhood on plane $\mathbf{Z}_i^2((k,m,n)^\top)$ for $i = 1, 2, 3$ [1]. Equation (9) implies that the 6-neighborhood is decomposed into mutually orthogonal three 4-neighborhoods [1].

A pair of points $(k,m,n)^\top$ and $x \in \mathbf{N}^6((k,m,n)\top)$ is a unit line segment in $\mathbf{Z}^3$, Furthermore, 6-connected four points which form a circle define a unit plane segment in $\mathbf{Z}^3$ with respect to the 6-connectivity. Therefore, we assume that our object is a complex of $2 \times 2 \times 2$ cubes which share at least one face each other. Thus, the surface of an object is a collection of unit squares which are parallel to planes $x = 0$, $y = 0$, and $z = 0$.

3 Curvature Indexes of Points

3.1 Planar Curvature Indexes

Since we are concerned with a binary discrete object, we affix 0 and 1 to points in the background and in objects, respectively. On $\mathbf{Z}^2$ three types of configurations of points which are illustrated in figure 1 exist in the neighborhood of a point $\times$ on the boundary. In figure 1, $\bullet$, and $\circ$, are points on the boundary and in the background, respectively. Setting $f_i \in \{0, 1\}$ to be the value of point x_i such that

$$x_3 = (m, n+1)^\top$$
$$x_5 = (m-1, n)^\top \quad x_0 = (m, n)^\top \quad x_1 = (m+1, n)^\top \tag{10}$$
$$x_7 = (m, n-1)^\top$$

the curvature of point x_0 is defined by

$$r(x_0) = 1 - \frac{1}{2} \sum_{k \in N} f_k + \frac{1}{4} \sum_{k \in N} f_k f_{k+1} f_{k+2}, \tag{11}$$

where $N = \{1, 3, 5, 7\}$ and $k + 8 = k$ [9]. The curvature indexes of configurations (a), (b) and (c) are positive, zero, and negative, respectively. Therefore, we call these configurations convex, flat, and concave, respectively, and affix the indexes $+$, 0, and $-$, respectively.

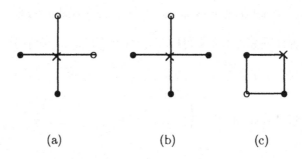

(a)   (b)   (c)

Fig. 1. Configurations of points on the boundary.

Setting n_+ and n_- to be the numbers of positive and negative points on a plane, respectively, we obtain the following theorem.

Theorem 1. *Let g to be the number of holes of a planar discrete object. If a discrete object is 4-connected, the relation*

$$n_+ - n_- = 4(1 - g) \tag{12}$$

is held.

Proof. If an object has no hole, it is possible to deform a 4-connected object to a rectangle. A rectangle implies the relation $n_+ = 4$. Thus, an object which has no hole holds the relation. Next, a hole is also deformable to a rectangle in the background. A rectangular hole holds the relation $n_- = 4$. Therefore, an object which has a hole holds the relation. Assuming that an object which has g holes holds the relation, an object which has $g + 1$ holes preserves the relation since the new hole is deformable to a rectangular hole. This concludes that n_- becomes $n_- + 4$, if a hole appears. Therefore, a 4-connected planar object holds eq. (12). $\square$

There exist 4 congruent patterns for patterns (a) and (c) in figure 1. These 8 patterns are independent. This theorem automatically derives an algorithm for the computation of the Euler characteristic of a planar discrete object using a table with 8 entries of (3×3)-local patterns, since points the vertex angles of which are 0-s do not affect to the Euler characteristic.

Step 1 Extract the boundary of object **O**.
Step 2 Count n_+ and n_- using the boundary following algorithm.
Step 3 Compute $n_+ - n_-$.

It is also possible to apply parallelly these 8 patterns to masking and matching.

3.2 Spatial Curvature Indexes

Using combinations of the planar curvature indexes on mutually orthogonal three planes which pass through a point x_0, we define the curvature index of a point x_0 in $\mathbf{Z}^3$ since the 6-neighborhood is decomposed into three 4-neighborhoods. Setting α_i to be the planar curvature index on plane $\mathbf{Z}_i^2(x_0)$ for $i = 1, 2, 3$, the curvature index of a point in $\mathbf{Z}^3$ is a triplet of two-dimensional curvature indexes $(\alpha_1, \alpha_2, \alpha_3)$ such that $\alpha_i \in \{+, -, 0, \emptyset\}$. Here, if $\alpha_i = \emptyset$, the curvature index of a point on plane $\mathbf{Z}_i^2(x_0)$ is not defined. The curvature indexes hold the following theorem.

Theorem 2. *For the boundary points, seven configurations*

$$(+, +, +), (+, +, -), (+, 0, 0),$$
$$(0, 0, \emptyset),$$
$$(-, -, -), (+, -, -), (-, 0, 0)$$
(13)

and their permutations are possible.

Proof. From the definition of the curvature indexes for points on $\mathbf{Z}^2$ and the connectivity of points in $\mathbf{Z}^3$, obviously the configurations with the indexes $+$ and $-$ exist. Therefore, we prove that the configurations such that $(*, *, 0)$, and $(0, 0, \emptyset)$, where $* \in \{+, -, 0\}$, and their permutations exist. From the configurations of points in a $3 \times 3 \times 3$ cube, configurations with 0-curvature on a plane exist. This geometric property automatically concludes that one of $*$ is zero. Thus, the curvature index is $(*, 0, 0)$ for $* \in \{+, -, \emptyset\}$. Furthermore, a congruence transform in $\mathbf{Z}^3$ yields a permutation of a code. □

4 Curvature and Spatial Vertex Angle

4.1 Local Configurations on the Boundary

Since a triplet of mutually orthogonal planes separates a space into eight parts, we call a one eighth of space an octspace. The number of octspaces determines

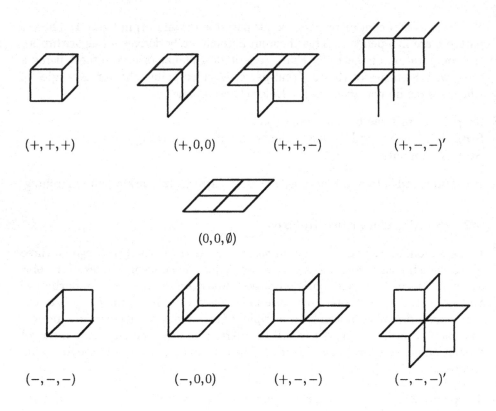

$(+,+,+)$ $(+,0,0)$ $(+,+,-)$ $(+,-,-)'$

$(0,0,\emptyset)$

$(-,-,-)$ $(-,0,0)$ $(+,-,-)$ $(-,-,-)'$

Fig. 2. Angles and configurations on the boundary.

configurations of points in a $3 \times 3 \times 3$ cube. There exist nine configurations in the $3 \times 3 \times 3$ neighborhood of a point on the boundary since these configurations separate $\mathbf{Z}^3$ to two parts which do not share any common points. These configurations are the same things introduced by Françon [13] for the analysis of discrete planes. The curvature analysis of discrete surfaces also derives these configurations.

For a spatial curvature index α, setting $n(\alpha)$ to be the number of octspaces in the $3 \times 3 \times 3$ neighborhood of a point, the relationships

$$n((+,+,+)) = 1, \quad n((+,0,0)) = 2, \, n((+,+,-)) = 3,$$
$$n((0,0,\emptyset)) = 4, \tag{14}$$
$$n((+,-,-)) = 4, 5, \, n((-,0,0)) = 6, \, n((-,-,-)) = 4, 7,$$

are held. From the decomposition property of the neighborhood and configurations of points in a 3×3 neighborhood on a plane, there exist two configurations for codes $(-,-,-)$ and $(+,-,-)$. Therefore, we set

$$n((-,-,-)) = 7, \, n((-,-,-)') = 4,$$
$$n((+,-,-)) = 5, \, n((+,-,-)') = 4. \tag{15}$$

4.2 The Vertex Angles

The curvature indexes of eq. (13) correspond to nine configurations of points which are illustrated in figure 2. Since $\frac{i}{8} = 1 - \frac{8-i}{8}$, we define the vertex angles of configurations as

$$\gamma((+,+,+)) = \frac{1}{8}, \quad \gamma((+,0,0)) = \frac{2}{8}, \, \gamma((+,+,-)) = \frac{3}{8},$$

$$\gamma((+,-,-)') = \frac{4}{8}, \quad \gamma((0,0,\emptyset)) = 0, \quad \gamma((-,-,-)') = \frac{-4}{8}, \tag{16}$$

$$\gamma((+,-,-)) = \frac{-3}{8}, \, \gamma(-,0,0) = \frac{-2}{8}, \, \gamma((-,-,-)) = \frac{-1}{8}.$$

Considering congruent transformations in $\mathbf{Z}^3$, since a code uniquely corresponds to a configuration, the vertex angle for a point on a surface is an extension of the chain code for a point on a curve.

Next, we clarify the geometric properties of nine configurations of points which are illustrated in figure 2. A configuration the vertex angle of which is 0 is flat in a $3 \times 3 \times 3$ local region.

A point the vertex angle of which is $\frac{1}{8}$ touches a plane the normal vector of which is one of $(\pm 1, \pm 1, \pm 1)^\top$, and a point the vertex angle of which is $\frac{-1}{8}$ also touches a plane the normal vector of which is one of $(\pm 1, \pm 1, \pm 1)^\top$. Therefore, these points are elliptic points in $\mathbf{Z}^3$.

A point the vertex angle of which is $\frac{2}{8}$ includes a line segment which is parallel to one of three axis of the coordinate system on a plane the normal vector of which is one of $(\pm 1, \pm 1, 0)^\top$, $(\pm 1, 0, \pm 1)^\top$, and $(0, \pm 1, \pm 1)^\top$. Furthermore, a point the vertex angle of which is $\frac{-2}{8}$ also includes a line segment which is parallel to one of three axis of the coordinate system on a plane the normal vector of which is one of $(\pm 1, \pm 1, 0)^\top$, $(\pm 1, 0, \pm 1)^\top$ and $(0, \pm 1, \pm 1)^\top$. Therefore, these points are parabolic points in $\mathbf{Z}^3$.

Points in the $3 \times 3 \times 3$ neighborhood of a point the vertex angle of which is $\frac{3}{8}$ exists in a half space separated by a plane the normal vector of which is one of $(\pm 1, 0, 0)^\top$, $(0, \pm 1, 0)^\top$, and $(0, 0, \pm 1)^\top$. Furthermore, points in the $3 \times 3 \times 3$ neighborhood of a point the vertex angle of which is $-\frac{3}{8}$ exists in a half space separated by a plane the normal vector of which is one of $(\pm 1, 0, 0)^\top$, $(0, \pm 1, 0)^\top$, and $(0, 0, \pm 1)^\top$. Therefore, these points are discrete hyperbolic-parabolic points in $\mathbf{Z}^3$.

Points in the $3 \times 3 \times 3$ neighborhood of a point the vertex angle of which is $\frac{4}{8}$ never exists in a half space separated by a plane the normal vector of which is one of $(\pm 1, 0, 0)^\top$, $(0, \pm 1, 0)^\top$ $(0, 0, \pm 1)^\top$, $(\pm 1, \pm 1, 0)^\top$, $(\pm 1, 0, \pm 1)^\top$, $(0, \pm 1, \pm 1)^\top$, and $(\pm 1, \pm 1, \pm 1)^\top$. Furthermore, points in the $3 \times 3 \times 3$ neighborhood of a point the vertex angle of which is $\frac{-4}{8}$ never exists in a half space separated by a plane the normal vector of which is one of $(\pm 1, 0, 0)^\top$, $(0, \pm 1, 0)^\top$ $(0, 0, \pm 1)^\top$, $(\pm 1, \pm 1, 0)^\top$, $(\pm 1, 0, \pm 1)^\top$ $(0, \pm 1, \pm 1)^\top$, and $(\pm 1, \pm 1, \pm 1)^\top$. These points are hyperbolic points in $\mathbf{Z}^3$.

5   The Euler Characteristic of Discrete Object

In this section, we prove the relation between the number of points n_i the vertex angles of which are $\frac{i}{8}$ and the number of tunnels of a discrete object.

5.1   The Euler Equation

Setting

$$n = (n_{-4}, n_{-3}, n_{-2}, n_{-1}, n_0, n_1, n_2, n_3, n_4)^\top \tag{17}$$

and

$$a = (-2, -1, 0, 1, 0, 1, 0, -1, -2)^\top, \tag{18}$$

we obtain the following theorems.

Theorem 3 . *For an object which has no tunnel, an object holds the relationship*

$$a^\top n = 8. \tag{19}$$

Theorem 4 . *For an object with g tunnels, the object holds the relationship*

$$a^\top n = 8(1 - g). \tag{20}$$

Theorem 5 . *Setting g, c, and p to be the number of tunnels, holes, and poles which are tunnels in holes, respectively, an object holds the relationship*

$$a^\top n = 8(1 - g + c - p). \tag{21}$$

There exist 8 congruent patterns for 6 patterns which affect to the Euler characteristic. These 48 patterns are independent. These theorems automatically derive an algorithm for the computation of the Euler characteristic of a discrete object using a table with 48 entries of $(3 \times 3 \times 3)$-local patterns, since points the vertex angles of which are 0-s and $\frac{\pm 2}{8}$-s do not affect to the Euler characteristic. The properties of indexes derive the following simple framework for the computation of the Euler characteristic.

Step 1 Decompose object **O** to collections of slices of planar patterns

$$\{O_{1\alpha}\}_{\alpha=1}^k, \ \{O_{2\beta}\}_{\beta=1}^m, \ \{O_{3\gamma}\}_{\gamma=1}^n,$$

which are perpendicular to $(1, 0, 0)^\top$, $(0, 1, 0)^\top$, and $(0, 0, 1)^\top$, respectively.
Step 2 Compute $r(x)$ for points on the boundaries of $O_{1\alpha}$, $O_{2\beta}$, and $O_{3\gamma}$.
Step 3 Compute γ for points $r(x) \neq 0$, according to the configurations.
Step 4 Count n_i for $i = \pm 1, \pm 3, \pm 4$.
Step 5 Compute $a^\top n$.

It is also possible to apply parallelly these 48 patterns to masking and matching.

5.2   Proofs of Theorems

Proof of Theorem 3 Since an object the vertex angles of which are 0-s, $\frac{1}{8}$-s, and $\frac{2}{8}$-s is a parallelopipedon, it holds eq. (19). Furthermore, since an object the vertex angles of which are 0-s $\frac{1}{8}$-s, $\frac{\pm 2}{8}$-s, and $\frac{\pm 3}{8}$-s is a collection of an appropriate number of parallelopipedons which contain points the vertex angles of which are $\frac{1}{8}$-s and $\frac{2}{8}$-s. Although the decomposition of an object into parallelopipedons is not uniquely determined, the decomposition is always possible. In the construction of an object from parallelopipedons, if k_1 points the vertex angles of which are $\frac{1}{8}$-s change k_3 points the vertex angles of which are $\frac{3}{8}$-s and k_{-3} points the vertex angles of which are $\frac{-3}{8}$-s, configurations of points in a $3 \times 3 \times 3$ cube imply the relationship $k_1 = k_3 + k_{-3}$. These geometric properties conclude that eq. (19) holds for all objects the vertex angles of which are 0-s $\frac{1}{8}$-s, $\frac{\pm 2}{8}$-s, and $\frac{\pm 3}{8}$-s.

Next for an object the vertex angles of which are 0-s, $\frac{\pm 1}{8}$-s, $\frac{\pm 2}{8}$-s, and $\frac{\pm 3}{8}$-s, it is possible to decompose an object into a collection of objects which do not contain points the vertex angles of which are $\frac{-1}{8}$-s. Furthermore, in the construction process, a point the vertex angle of which is $\frac{1}{8}$ and a point the vertex angle of which is $\frac{-2}{8}$ forms a point the vertex angle of which is $\frac{-1}{8}$. These geometric properties also conclude that eq. (19) holds for an object the vertex angles of which are 0-s $\frac{\pm 1}{8}$-s, $\frac{\pm 2}{8}$-s, and $\frac{\pm 3}{8}$-s.

Finally, it is possible to decompose an object which contains points the vertex angles of which are $\frac{4}{8}$-s and $\frac{-4}{8}$-s into a collection of objects which do not contain points the vertex angles of which are $\frac{\pm 4}{8}$-s. Conversely, combining a pair of objects the vertex angles of which are $\frac{1}{8}$-s and $\frac{2}{8}$-s, it is possible to construct an object which contains points the vertex angles of which are $\frac{4}{8}$-s and $\frac{-4}{8}$-s. These geometric properties also conclude that eq. (19) holds for an object the vertex angles of which are 0-s, $\frac{\pm 1}{8}$-s, $\frac{\pm 2}{8}$-s, $\frac{\pm 3}{8}$-s, and $\frac{\pm 4}{8}$-s. Thus, eq. (19) holds for objects without tunnels.

Proof of Theorem 4 From theorem 3 all discrete objects without tunnels are topologically equivalent to an object which has four corners the vertex angles of which are $\frac{1}{8}$-s. These objects are parallelopipedons, which are topologically equivalent to a cube which is a sphere in the 6-connected discrete space.

Considering an object with a tunnel, it is possible to eliminate a tunnel using a wall which is parallel to one of planes $\mathbf{Z}_i^2(x_0)$ for $i = 1, 2, 3$ [4]. Thus, a tunnel is transformed to a pair of wells. Since an object with a pair of wells is an object without tunnels, it is possible to deform this object to an object with four vertices the vertex angles of which are $\frac{1}{8}$-s, eight vertices the vertex angles of which are $\frac{-1}{8}$-s, eight vertices the vertex angles of which are $\frac{-3}{8}$-s. If we eliminate a wall which separates a pair of wells, we obtain a tunnel, and eight vertices the vertex angles of which are $\frac{-1}{8}$-s disappear. Thus, for an object with a tunnel, eq. (20) holds.

Assuming eq. (20) for an object with $(p - 1)$ tunnels, for an object with p tunnels, it is possible to operate the same procedure in order to transform a tunnel to a pair of wells. Furthermore, it is also possible to deform a pair of wells to a pair of wells which consist form eight vertices the vertex angle of which are $\frac{-1}{8}$-s, and eight vertices the vertex angles of which are $\frac{-3}{8}$-s. If we eliminate a tunnel which separates a pair of wells, we obtain a tunnel, and eight vertices the vertex angles of which are $\frac{-1}{8}$-s disappear. Thus, for an object with p tunnels, eq. (20) holds.

Proof of Theorem 5 The vertexes of the inner surface holds eq. (19), because the inner surface has the same vertex angles with the vertex angles of the outer surfaces. Since the Euler characteristic of an object with holes is the sum of the Euler characteristics of the outer boundary and the inner boundaries, we have

$$a^\top n = 8(1 - g) + \sum_{k=1}^{c} 8(1 - p_k), \qquad (22)$$

where c is the number of holes and p_i is the number of poles in the i-th hole.

6 Conclusions

We introduced the curvature indexes of the boundary of a discrete object, and using these indexes of points, we defined the vertex angles of discrete surfaces as an extension of the chain codes of digital curves. Furthermore, we proved the relation between point configurations and the genus of a discrete object. This is the angular Euler characteristic of a discrete object. This relation derives a parallel algorithm for the computation of the Euler characteristic of a discrete object. In this paper we assumed the 6-connectivity for points in $\mathbf{Z}^3$. However, extensions of the angular Euler characteristics to objects which defined using 18- and 26- connectivities are possible if we define the curvature indexes for these connectivities.

In a series of papers [16] [17] [18] Bieri and Nef proposed sweeping algorithms for the computation of geometric properties of a polytope. The main idea of them is the sweeping of a space by a hyperplane. This idea is equivalent to the decomposition of a space to collections of lower dimensional spaces. This idea goes back to the Radon transform, which is now the theoretical background of computerized tomography [19] [20]. Our method is considered as a discrete version of the space-sweeping algorithm. Our method, however, requires many predetermined planes.

Regular 3-graphs enjoy nice properties for proving the combinatorial characteristics of polyhedrons [14]. The surface of a 6-connected object defines a graph the degrees of nodes of which run from 3 to 6, if we consider points and bonds as nodes and edges, respectively. In the proof of theorem 3, we dealt with an object the vertex angles of which are 0-s, $\pm\frac{1}{8}$-s, $\pm\frac{2}{8}$-s, and $\pm\frac{3}{8}$-s. For these graphs, if we first eliminate all bonds connecting points the vertex angles of which are 0-s, second, remove all isolated nodes, and third change all nodes the vertex angle of which are $\pm\frac{2}{8}$-s to edges, we obtain a 3-graph. In the proof of theorem 4, we changed an object which contains points the vertex angles of which are $\pm\frac{4}{8}$-s to an object without them. Furthermore, in the proof of theorem 5, we used this methodology. Therefore, the proofs provide us rewriting rules from a graph defined by the connectivity of points on the boundary of a discrete object to a 3-graph.

The decomposition of an object into complexes derives the definition of the computation method for the Euler characteristic of a digital object, that is, the total number of the all possible simplexes in an object defines the Euler characteristic. In reference [10], Toriwaki and coworkers defined the topological properties of discrete set using all points in the region of interest [11], and applied this idea for the computation of the Euler characteristic of a discrete binary object. Furthermore, they derived a logic function for the computation of the Euler characteristic. Their method requires all points in the region of interest. Our method, however, requires points on the surface which is extracted using an appropriate method [1] [2] [3]. It is possible to define topological characteristics of objects using only points on the boundary [5]. Our method is a discrete version of this idea. From computational view points, the later method requires less amounts of the memories.

References

1. Imiya, A.: Geometry of three-dimensional neighborhood and its applications, Transactions of Information Processing Society of Japan, Vol. 34, pp. 2153-2164, 1993 (in Japanese).
2. Kenmochi, Y., Imiya, A., and Ezuquera, R.: Polyhedra generation from lattice points, Miguet, S., and Montanvert, A., and Ubéda, S. eds.: *Discrete Geometry for Computer Imagery*, pp.127-138, Lecture Notes in Computer Science, Vol. 1176, Springer-Verlag: Berlin, 1996.
3. Kenmochi, Y. and Imiya, A.: Deformation of discrete object surfaces, Sommer, G., Daniilidis, K. and Pauli, J. eds.: *Computer Analysis of Images and Patterns,* pp. 146-153, Lecture Notes in Computer Science, Vol. 1296, Springer-Verlag: Berlin, 1997.
4. Aktouf, Z., Bertrand, G., and Perrton, L.: A three-dimensional holes closing algorithm, pp. 36-38, Miguet, S., and Montanvert, A., and Ubéda, S. eds.: *Discrete Geometry for Computer Imagery,* Lecture Notes in Computer Science, Vol. 1176, Springer-Verlag: Berlin, 1996.
5. Grünbaum, B.:Convex Polytopes, Interscience: London, 1967.
6. Okabe, A., Boots, B., and Sugihara, K.: *Spatial Tessellations Concepts and Applications of Voronoi Diagrams,* John Wiley& Sons: Chichster, 1992.
7. Hopf, H.:*Differential Geometry in the Large,* Lecture Notes in Mathematics, Vol. 1000, Springer-Verlag: Berlin, 1983.
8. Serra, J.:*Image Analysis and Mathematical Morphology,* Academic Press: London, 1982.
9. Toriwaki, J.-I.: *Digital Image processing for Image Understanding,* Vols.1 and 2, Syokodo: Tokyo, 1988 (in Japanese).
10. Yonekura, T., Toriwaki, J.-I., and Fukumura, T., and Yokoi, S.: On connectivity and the Euler number of three-dimensional digitized binary picture, Transactions on IECE Japan, Vol. E63, pp. 815-816, 1980.

11. Toriwaki, J.-I., Yokoi, S., Yonekura, T., and Fukumura, T.: Topological properties and topological preserving transformation of a three-dimensional binary picture, Proc. of the 6th ICPR, pp. 414-419, 1982.
12. Lee, C.-N., Poston, T., and Rosenfeld, A.: Holes and genus of 2D and 3D digital images, CVGIP Graphical Models and Image Processing, Vol. 55, pp. 20-47, 1993.
13. Françon, J.: Sur la topologie d'un plan arithmétique, Theoretical Computer Sciences, Vol. 156, pp. 159-176, 1996.
14. Bonnington, C. P. and Little, C.H.C.: *The Foundation of Topological Graph Theory* Springer-Verlag: Berlin, 1995.
15. Bieri, H. and Nef, W.: Algorithms for the Euler characteristic and related additive functionals of digital objects, Computer Vision, Graphics, and Image Processing Vol. 28, pp. 166-175, 1984.
16. Bieri, H. and Nef, W.: A recursive sweep-plane algorithm, determining all cells of a finite division of R^d, Computing Vol. 28, pp. 189-198, 1982.
17. Bieri, H. and Nef, W.: A sweep-plane algorithm for computing the volume of polyhedra represented in Boolean form, Linear Algebra and Its Applications, Vol. 52/53, pp. 69-97, 1983.
18. Bieri, H. and Nef, W.: A sweep-plane algorithm for computing the Euler-characteristic of polyhedra represented in Boolean form, Computing, Vol. 34, pp. 287-304, 1985.
19. Stark, H. and Peng, H.: Shape estimation in computer tomography from minimal data, Proceedings of 9th ICPR, Vol. 1, pp. 184-186, 1988.
20. Ludwing, D.: The Radon transform on Euclidean space, Comm. Pure and Applied Mathematics, Vol. 19, pp. 49-81, 1966.

Fast Estimation of Mean Curvature on the Surface of a 3D Discrete Object

Alexandre LENOIR

GREYC UPRESA 6072
ISMRA 6 Bd. du Maréchal Juin 14050 CAEN Cedex, France.
Email : Alexandre.Lenoir@greyc.ismra.fr

Abstract. Discrete surfaces composed of surfels (a surfel is a facet of a voxel) have interesting features. They represent the border of a discrete object and possess the Jordan property. These surfaces, although discrete by nature, represent most of the time real world continuous surfaces for which local geometrical characteristics are useful for registration, segmentation, recognition and measure. We propose a technique designed to estimate the mean curvature field of such a surface. Our approach depends on a scale parameter and has a low computational complexity. It is evaluated on synthetic surfaces, and an application is presented : the extraction of sulci on a brain surface.
Index terms – Discrete surfaces, surfels, geometric invariant, mean curvature

1 Introduction

Modern imaging techniques like MRI or confocal microscopy produce 3D digital images that represent real world scenes. A segmentation step followed by a labelling of the resulting binary image yields well identified 3D discrete objects. Their surface contains a lot of useful information. Second order differential invariants such as curvatures play a key role in image processing. For example, they are used for registration of 3D images, or for representation, recognition and segmentation of objects, as presented in [4, 7, 11]. They may also be used as internal forces that drive the deformation of an active surface. Mean curvature has a special physical signification : zero mean curvature surfaces are called minimal surfaces, because given a closed curve in 3D-space, the surface of minimal area that has this curve for border has everywhere a zero mean curvature [5] p.199. Soap films are familiar examples of minimal surfaces.

1.1 Notions of discrete surfaces and related algorithms

There are several definitions of discrete surfaces. Some are composed of subsets of voxels, seen as points of $\mathbb{Z}^3$ that verify some special topological properties [15]. Their definition is still studied in order to adapt it to intuitive notions of surfaces [3, 12]. Some other are made of surfels, that can be seen as the facets of voxels, elementary cubes whose centre has integer coordinates. This context is called the cuberille model and we will use this notion in the following. A formal description in arbitrary dimension of this model can be found in [8] and [19], and will partially be recalled in section 2. This kind of surface is still actively studied in arbitrary dimension, but their detection and properties are already well known in 3D [18,20]. Efficient techniques that give the volume enclosed by these surfaces exist [18]. Furthermore, one may also find boundary based techniques that produce the geometrical moments of an homogenous volume enclosed by the surface in [22]. Many papers have been written to propose and to justify methods that detect the connected components of surfaces in binary scenes [17,20]. However, specific techniques that estimate shape information like normal vector field, area or curvature for this surface representation seem to lack. A first step has been proposed in [10]. We use the word estimate, because from now, we will suppose that the discrete surface actually represents a smooth continuous surface, and we are interested in obtaining the geometric properties of this continuous surface. The computation of geometrical properties of the discrete surface itself is straightforward but meaningless. Moreover, as the surface of an object of practical interest is made of a big number of surfels (typically from 50 000 to 400 000), an efficient approach is welcome to avoid long computation time and to allow interactive use or multiscale analysis.

1.2 Estimation of geometric invariants

We will now briefly survey the methods that may be found in the literature and aimed at estimating the differential properties of discrete or sampled surfaces. In 2D, the estimation of differential invariants of curves like tangent, length and curvature was already well studied (see [21] and its references). The main problem of the estimation of parameters on a discrete surface (3D case) is that no parametrization of the surface exists contrarily to the 2D case. But for 2D1/2 surfaces (range images) represented by a depth map, a natural coordinate system is the 2D position. On this kind of images, one can get a local model, either by using a polynomial fit like in [4] or by computing partial derivatives of the depth function along coordinate lines. Classical formulas of differential geometry are then used to get the invariants from the partial depth derivatives or from the fitted polynomial. Different approaches exist for true 3D surface analysis. THose presented in [16] and [14] use partial derivatives of the 3D grey level image on the surface trace to compute the differential properties of the surface. These methods need to scan the entire 3D volume several times. A method presented in [13] uses the fitting of a quadratic surface patch on the surface trace. This method is to be more computationally efficient since it works only on the surface trace.

1.3 Our approach

The method described here uses the regular structure of the discrete surface as the support of functions of vectorial values that describe the geometry of the surface at the surfel scale. Then, we convolve recursively these values by a low pass filter along special curves, in order to get a regional average of these local geometrical values. These curves define coordinates curves that are common to several surfels. They are defined globally, and not locally. As the surfel belongs to two of these curves, the curves define a local parametrization of the surface. The geometric invariants are then obtained via local calculation. We use this uniform framework to compute both normals and mean curvature field. Note that the normals field is needed to get mean curvature. One of the strength of this method is its time complexity in $O(\sigma.\sqrt{(n)} + n)$, where σ is the scale parameter of the calculation, and n the number of surfels of the object. We can see that it is only slightly dependent on the scale parameter. This is be a attractive in a multiscale context. Another advantage is the simplicity and the uniformity of tools used to compute both normals and curvatures. The width of the neighbourhood taken into account is directly controlled by the scale parameter.

1.4 Outline of the paper

This paper is organised in the following way : in section 2 we give the basic definitions relative to discrete surfaces. Then, we explain the recursive computation of the convolution product of a summable function and a periodic function. In section 3 some basic notions of differential geometry of parametrized curves and surfaces are recalled. In section 5 we explain the calculus of mean curvature which implementation is detailed in section 4. The computational complexity is estimated in section 6 and experimental results on families of analytic objects are presented and discussed in section 7.

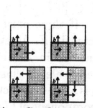 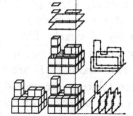 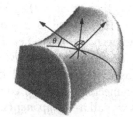

(a)   Configurations of voxels and bels that meet the bel b at the edgel e. Grey squares are $1-$ voxels, white one are $0-$voxels

(b) Slices and slice contours of an object along the three families of main planes. Arrows show the orientation of the contours

(c)   A surface, its normal, principal curvature directions and a curve of the surface with an angle of θ with the main curvature direction.

Fig. 1. Special curves on discrete and continuous surfaces

2 Notations and basic definitions

We first describe the nature of the discrete we use, as well as the curves we consider on it. Afterwards we explain the main calculation tool used in our method: the recursive computation of a convolution product.

2.1 Surfaces of 3D discrete objects

These definitions are mainly drawn from [8] and [19] but we restrict ourselves to the 3D case.

Main vectors of $\mathbb{R}^3$, Δ, operations involving these vectors. We consider the euclidian vectorial space $\mathbb{R}^3$ and a direct orthonormal basis (O, w_1, w_2, w_3). We will use the sets of indices $I_+^3 = \{1, 2, 3\}$, $I^3 = \{-3, -2, -1, 0, 1, 2, 3\}$, and $I_*^3 = I^3 \setminus \{0\}$. We define $\forall i \in I_+^3, \delta_i = \omega_i, \delta{-i} = -\delta_i$, the null vector δ_0, $\Delta = \{\delta_i / i \in I^3\}$ and $\Delta_* = \{\delta_i / i \in I_*^3\}$. We define the multiplicative operator $\otimes$ from the vector cross product $\wedge$: $\otimes : \begin{vmatrix} I^3 \times I^3 \to I^3 \\ (x,y) \to z = x \otimes y / \delta_z = \delta_x \wedge \delta_y \end{vmatrix}$.

Voxel, binary scene. Let $x \in \mathbb{R}^3$. We denote x_i the ith coordinate of x. $\mathbb{R}^3$ is divided into unit cubes called *voxels* by a set of planes orthogonal to the axes $\{P_{i,j} \in I_+^3, j \in \mathbb{Z} + \frac{1}{2} \text{ and } P_{i,j} = \{x \in \mathbb{R}^3 / x_i = j\}\}$. We identify each voxel with its central point of $\mathbb{Z}^3$. A binary scene of $\mathbb{Z}^3$ is a function $B : \mathbb{Z}^3 \to \{0, 1\}$. We call $B(v)$ the value of the voxel v. We note $1(B) = B^{-1}(1)$ and $0(B) = B^{-1}(0)$.

n−adjacency, n−path, n−object, n−background. Two voxels c and d are said to be 6-adjacent if they share a face. They are 18−connected if they are 6-adjacent or if they share an edge. For $n \in \{6, 18\}$, an n−*path* of lenght l $p = [v_0, \ldots, v_l]$ is a sequence of $l + 1$ voxels so that $\forall j \in [0, l - 1]$ v_j is n−adjacent to v_{j+1}. Let E be a set of voxels. Let $x, y \in E$. If there exists a n−path from x to y in E, we say that x and y are n−connected in E. A set E of voxels is n−connected iff any two voxels of E are n−connected in E. Since n−connectedness is an equivalence relation, we consider its equivalence classes that we call n−*components*. If B is a binary scene, an n−*object* is an n−connected component of $1(B)$ and an n−*background* is an n−connected component of $0(B)$.

Surfel, Surfel type, Surface, Boundary, Border, Bel. A *surfel* is an oriented surface element that is identified by the pair (v_1, v_2) of 6-adjacent voxels, of which it is the common face. Therefore, the vector $\delta = v_2 - v_1 = \delta_i \in \Delta_*$ can take six distinct values. The *type of the surfel* s is $T(s) = i$. We will call v_1 the start voxel of s, and v_2 its end voxel. A *surface* is a set of surfels. The boundary of two disjoint sets of voxels E_1 and E_2 is the set $\beta(E_1, E_2) = \{s = (a_1, a_2)/s \in \Sigma$ and $a_1 \in E_1$ and $a_2 \in E_2\}$. A *bel* of a binary scene B is a surfel $a = (c, d)$ so that $c \in 1(B)$ and $d \in 0(B)$. The boundary of a binary scene is the set of its bels. A

$\kappa\lambda$−border is the boundary of a κ−object and of a λ−background. In the sequel we will consider $\kappa\lambda$−borders, where $\kappa = 6$ and $\lambda = 18$. Such a border is proved to be connected in some sense, to have a connected interior and exterior and to possess the Jordan property.

Edgel, Edgel's type, Bel adjacency, Surface graph. We call *edgel* of the surfel a the pair $e = (a, \delta_i)$, with $\delta_i \in \Delta_*$ and $|i| \neq |T(a)|$. $T(e) = i$ is the type of e. We also say that the surfel meets the surfel at the edgel iff a and a' share an edge. We can define an adjacency relation between bels such as each edgel corresponds to a single adjacent bel. This neighbouring relation define the notion of path on a surface and of connected component of a surface, as well as the notion of surface graph. Each surfel has exactly four neighbours (one per edgel).

Slices, Slice contour, Slice contour function. A slice of $\mathbb{Z}^3$ is a set of voxels in which one coordinate is fixed, the two others being free. The slice denoted $Sl_{i,j}$ is the set of voxels whose ith coordinate is j. Let be $b = (c, d)$ a bel of type i. It belongs exactly to two *slice contours* denoted by $SlC_{i,j}, i \in I_3^+ \setminus \{|T(b)|\}$. We call i the type of the slice contour $SlC_{i,b}$. These slice contours are images of slice contour functions, denoted $SlCF_{i,b}$. The succession of slice contour bels is naturally defined by their adjacency relation and their type :
$SlCF_{i,b} : \begin{vmatrix} \mathbb{Z} \to B \\ z \to SlCF_{i,b}(z) \end{vmatrix}$ with $SlCF_{i,b}(0) = b, \forall z \in \mathbb{Z}, t = T(z),$ and $e = \delta_i \wedge \omega_i$.
$SlCF_{i,b}$ is then recursively defined in the following way :
$$\begin{cases} SlCF_{i,b}(z + 1) \text{ is the adjacent bel of } SlCF_{i,b}(z) \text{ at } e \\ SlCF_{i,b}(z - 1) \text{ is the adjacent bel of } SlCF_{i,b}(z) \text{ at } -e \end{cases}$$
Slice contours of finite objects are periodic lists of adjacent bels whose start voxels are in a same plane. A slice may contain several slice contours as shows Fig. 1(b) for horizontal slices.

2.2 Recursive computation of a discrete convolution product

Let $g : \mathbb{Z} \to \mathbb{R}$ be a convolution kernel. Let f be a periodic function of $\mathbb{Z}$ with values in $\mathbb{R}$. We want $\chi = g * f$. In [6] is explained how to approximate a gaussian kernel and its derivatives by a sum of exponential as well as the recursive implementation of a non causal filter. The k−order recursive calculation of χ is implemented by splitting g into a sum of two functions g_- et g_+, respectively null upon $\mathbb{R}^+$, and $\mathbb{R}_*^-$. For each of these functions, we need only the k previous values of the convolution products as well as the current value of f and its $k - 1$ previous values. Notice that the method proposed depends on a scale parameter but that the design of the filter for any choice of scale parameter is very little time consuming. This is important since the period of the function we wish to convolve may be as small as 4 (the length of a slice contour bounding a single voxel). Whenever this case occurs it would be meaningless to keep a wide kernel for these short slice contours. In such a case, the scale of our filter is adapted so that it is always less than a fraction of the length of the slice contour it will convolve. This fraction is 4 in our implementation.

Contrary to the usual context of signal filtering, initial values of the convolution are not zero, because the function we want to convolve is not everywhere zero outside a given interval. We therefore need an initialisation step for the recursive calculation. At the start of the recursion, one can truncate g outside an appropriate interval and then compute the first k convolution products in a non recursive way. We normalise our filters such that $\sum_{x \in \mathbb{Z}} g(x) = 1$ for an averaging kernel and $\sum_{x \in \mathbb{Z}} x.g(x) = 1$ for a derivative kernel.

3 Local differential geometry

We have seen in section 2.1 that the nature of the surface defines naturally slice contours which are plane curves. If we suppose that the surface is a discretization of a smooth surface, then the slice contour is a discretized smooth plane curve. Consequently, we will recall the basic concepts of differential geometry of smooth parametrized plane curves and surfaces. These classical formulas of differential geometry of surfaces are drawn from [9].

3.1 Plane curves

Let P be a plane with a direct othonormal basis $(O; i, j)$. We set $k = i \wedge j$. Let $C : |^{U \subset R \to P}_{s \to C(s)}$ be a curve in this plane. We suppose that this curve is parametrized by its arc length:$\|\frac{\delta C}{\delta s}\| = 1$. The unit tangent vector of this curve is $\tau = \frac{\delta C}{\delta s}$ and the unit normal as $\nu = \tau \wedge k$. The algebraic curvature of the curve is $\rho = \frac{\delta \tau}{\delta s}$.

3.2 Surfaces

Here we consider a smooth surface S in the euclidian 3D space (see Fig. 1(c)). The surface is parametrized in a neighbourhood of a point p by $M : |^{U \subset \mathbb{R}^2 \to S}_{(u,v) \to M(u,v)}$ with $M(0,0) = p$. Let C_1 and C_2 be any two curves on the surface with $C_1(0) = C_2(0) = p$. Provided that the curves are not mutually parallel at p, the two tangents of the curves at p, $\tau_1(0)$ and $\tau_2(0)$ and are in a plane that is the tangent plane of the surface at p. N_p, the normal unit vector of T_p is colinear to $\tau_1(0) \wedge \tau_2(0)$. Let P_0 be a plane that contains N_p. Now let C_θ, $C_\theta(0) = p$ be a curve that lies both on the surface and on a plane P_θ that contains N_p. P_θ has a dihedral angle θ with P_0. Let $\rho(\theta)$ and $\tau(\theta)$ be the curvature and tangent of C_θ at p. $\rho(\theta)$ considered as of map from $[0, \pi[$ on $\mathbb{R}$ has in general two extremal values, $\lambda_1 = \rho(\theta_1)$ and $\lambda_2 = \rho(\theta_1 + \frac{\pi}{2})$. From now, we will assume that $\theta 1 = 0$. The two directions of extremal curvatures are the principal directions of the surface.

3.3 Curves on surfaces

Let C be a curve on a surface S. Let τ, ν and ρ be its tangent, normal and curvature at p. Then the curvature at p of a curve C' on S and on a plane containing N_p with the same tangent τ is $\rho_\nu = \rho.(N_p.\nu)$ and is called the normal curvature of C at p. Normal curvature depends only on the tangent of the curve and is noted $\rho(\theta)$ if $\tau = \tau(\theta)$.

4 Computing surface mean curvature from curve curvature

We explain here the principle of our method. We use slice contours and their curvatures to get the mean curvature of the surface. Slice contours are plane curves, and the computation of plane curves curvature is a common topic in the field of shape analysis. One can see [21] for a survey of curvature estimations methods. A surfel is crossed by two slice contours. We suppose we know their curvatures at each surfel. Provided the normal to the surface on the surfel s, we can have the normal curvature of each curve. We may then use the Euler formula that links the normal curvature c of a curve to the angle x of the tangent of the curve with one of the two main curvatures (λ_1 and λ_2) of the surface :

$$\text{Euler formula} \qquad c = \lambda_1.\cos(\theta)^2 + \lambda_2.\sin(\theta)^2 \qquad (1)$$

The mean curvature is $H = \frac{\lambda_1 + \lambda_2}{2}$, it is also equal to $\frac{c+c'}{2}$, when c and c' are the normal curvatures of two orthogonal curves that lie on the surface. If the curves are not orthogonal, as we do not know θ, we cannot get H. However, it is possible to get H if we know three curves $(C_i)_{i \in I_*^3}$ with normal curvatures $(c_i)_{i \in I_*^3}$ that pass through the same surface point p. First, we note that (1) can be rewritten $c = H + d.\cos(2.\theta)$ with $d = \frac{\lambda_1 - \lambda_2}{2}$. For $(i,j) \in I_*^3 \times I_*^3$, we note $\alpha_{i,j}$ the angle between C_i and C_j at p. We have the set of equations $\forall i \in I_*^3, c_i = H + d.\cos(2.(\theta + \alpha_{i,j}))$ where θ is the angle between C_i and the main curvature direction. We wish to find a combination of normal curvatures in order to eliminate d for any value of θ. We use the trigonometric relation:

$$
\begin{aligned}
& \cos(2.\theta).\sin(2.(\alpha_{1,3} - \alpha_{1,2})) \\
& + \cos(2.(\theta + \alpha_{1,2})) + \sin(-2.\alpha_{1,3}) \\
& + \cos(2.(\theta + \alpha_{1,3})) + \sin(2.\alpha_{1,2}) = 0
\end{aligned}
\qquad (2)
$$

Using the triangle relation $\alpha_{1,2} + \alpha_{1,2} + \alpha_{1,2} = \pi$, with $0 \le \alpha_{1,2}, \alpha_{2,3}, \alpha_{3,1} \le \frac{\pi}{2}$, we obtain $\alpha_{1,3} - \alpha_{1,2} = \alpha_{2,3} - \pi$. Then (2) can be rewritten:

$$
\begin{aligned}
& \cos(2.\theta).\sin(2.\alpha_{2,3}) \\
& + \cos(2.(\theta + \alpha_{1,2})).\sin(2.\alpha_{3,1}) \\
& + \cos(2.(\theta + \alpha_{1,3})).\sin(2.\alpha_{1,2}) = 0
\end{aligned}
\qquad (3)
$$

The coefficients of a linear combination appear clearly and are all positive. They do not depend on θ but only on the relative angles of the curves. We define $\beta_i = \sin(2.\alpha_{j,k})$ with $j \otimes = i \in I_*^3$ and $\alpha_i = \frac{\beta_i}{\sum_{i \in I_*^3} \beta_i}$. Finally we get

$$H = \sum_{i \in I_*^3} (\alpha_i.c_i) \qquad (4)$$

5 Application to surfaces of discrete objects

We will now apply the previouss onto discrete surfaces. Only two curves pass trough a surfel, but if we consider a neighbourhood of a surfel we can verify that one of the following cases is true:

- One or two of the components of the normal to the surface are near zero. There exist only two types of slice contours, but they are mutually orthogonal. So we can have mean curvature from curves curvatures.
- None of the components of the normal to the surface is near zero. Slice contours are not mutually orthogonal but there exist three types of them. We may apply (4) to get the mean curvature.

We suppose from here that the normal to the surface is (a, b, c), with $a^2+b^2+c^2 = 1$. The angles between slice contours may be expressed with the coordinates of the normal to the surface. Tangent vectors of slice contours are parallel to $(-b, a, 0)$ and $(-c, 0, a)$, for slice contours of type 2 and 3. We have $\cos(\alpha_{2,3}) = \frac{bc}{\sqrt{a^2+b^2}.\sqrt{a^2+c^2}}$, and $\sin(\alpha_{2,3}) = \frac{a}{\sqrt{a^2+b^2}.\sqrt{a^2+c^2}}$. Therefore we have :

$$\beta_1 = \sin(2.\alpha_{2,3}) = 2.\cos(\alpha_{2,3}).\sin(\alpha_{2,3}) = \frac{a.b.c}{(a^2 + b^2).(a^2 + b^2)} \qquad (5)$$

and by definition of α_1 :
$$\alpha_1 = \frac{b^2 + c^2}{2} = \frac{1 - a^2}{2} \qquad (6)$$

5.1 Normal estimation

The surface graph of the 18 or 6-connected object is first computed thanks to an adaptation of surface tracking algorithm (see [1]). The whole surface graph, and not only its vertices are needed. A transversal of this graph yields the length, as well as a start surfel of each slice contour. We explained in [10] the way we compute the normals, by estimating tangents and then by doing a vector cross product, and another smoothing in order to have a good result at each surfel.

5.2 Slice contour curvature estimation

We use the tangents of each of the curves computed previously. We first normalise them and then a derivative filter along the slice contours is applied. Note that derivation is done with respect to the discrete arc length, and not with respect to the continuous arc length like in the classical definition. But as we know the ratio between the smoothed discrete arc length and the continuous arc length is the euclidian norm of the non normalised tangents, we divide the result of the derived normalised tangents by this norm. We suppose now that in a neighbourhood of a surfel s, the mean curvature varies smoothly and that the slice contour curvatures depends on the type of the slice contour but varies also smoothly inside the neighbourhood. We will note ρ_i the curvature and c_1 the normal curvature of a slice contour C_i of type i in the neighbourhood of s. For instance, for a slice of type 1, we have $c_1 = \rho_1.\sqrt{(b^2 + c^2)}$.

5.3 Mean curvature estimation

Using (6) and the unit normal vector, we compute α_1 for each of the two slice contours of each surfel. We now assume that the ratio of surfels of type 2 and 3 on this slice contour are $r_{1,2}$ and $r_{1,3}$. These ratios may be estimated by a low pass filter along each curve. For a slice contour of type 1, we compute the following values : $\alpha_1.c_1 + \dfrac{\alpha_3.c_3}{r_{1,2}}$ or surfels of type 2 and $\alpha_1.c_1 + \dfrac{\alpha_2.c_2}{r_{1,3}}$ for surfels of type 3. Note that when we are on a surfel of type 2, $r_{1,2}$ cannot be zero. We recall that a surfel of type 2 intersect slice contours of types 1 and 3. Then we convolve this value with the same smoothing filter used to get the ratios and we get $\alpha_1.c_1 + \alpha_1.c_1 + \alpha_1.c_1 = H$, following (4).

6 Evaluation of algorithmic complexity

If the object is bounded by cube of size n, the number of slice contours is roughly $3.n$, and the number of surfels of its surface is around n^2 . We want to compute a field at the scale σ. The cost of the initialisation of the recursive calculation is proportional to the number of slice contours and to the scale : $O(\sigma.n)$. The recursive computation needs a constant computational amount for each surfel. There are then only local computations. As several convolutions are needed, we just have to add their costs. In conclusion, the cost of our method is $O(\sigma.n+n^2)$.

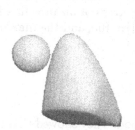
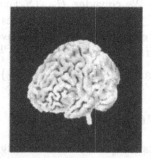

(a) A paraboloid and a sphere with identical curvatures at top.

(b) Mean curvature field on a brain surface

(c) After filtering the parts of the surface of negative mean curvature, we have extracted their centre lines

Fig. 2. Experiments and application to mean curvature field.

7 Experimental results and applications

We have made experiments in order to estimate the precision of our technique. All the experiments were done with an exponential filter with $\sigma = 4$ for averaging filters, and the derivative of an exponential filter with $\sigma = 2.5$. Filters with lower scale may have been used for short slice contours as explained in section 2.2.

7.1 Experimental results for spheres

Table 1. Experimental results for spheres

r	$\frac{1}{r}$	$H_{min}(S_r) - H_{max}(S_r)$	$\Delta_{max}(S_r)$	$\Delta_{max}(S_r)$
5	0.2	$0.248 - 0.261$	30.5%	5.9%
10	0.1	$0.106 - 0.107$	7%	5.85%
15	0.0667	$0.0678 - 0.0680$	2%	8.17%
20	0.05	$0.0500 - 0.0502$	0.4%	10.2%
30	0.0333	$0.0329 - 0.0330$	1.2%	12.4%
40	0.025	$0.0246 - 0.0246$	1.6%	16.1%

The mean curvature of a sphere is everywhere the inverse of its radius. We have done a set of experiment on spheres of different radius, digitized one hundred times at a random position. The results of these experiments are shown in table 7.1. For a given radius r, let S_r be the set of digitized spheres of radius r. For $S \in S_r$ we denote $s(S)$ the set of its surfels and $H(s)$ the mean curvature computed at the surfel s. We define $\overline{H}(S) = \frac{1}{|S|} \sum_{s \in S} H(s)$ the mean computed value over the surfels of S. We consider $H_{min}(S_r) = \min\{\overline{H}(S), H \in S_r\}$ and $H_{max}(S_r) = \max\{\overline{H}(S), H \in S_r\}$ the extremal values for all digitized spheres of a given radius of $\overline{H}(S)$. We computed also the relative errors of mean values for all spheres $\overline{\Delta}(S_r) = \max\{|H_{max}(S_r) - \frac{1}{r}|, |H_\in(S_r) - \frac{1}{r}|\}.r.100$, and the maximal relative error $\Delta_{max}(S_r) = \max\{\frac{\max\{|\overline{H}(S) - H(s)|, s \in S\}}{\overline{H}(s)}\}$.

7.2 Experimental results for paraboloids

We have seen that the results for spheres were rather good. They showed that the result is quite insensitive to orientation although the data structure itself is not isotropic at all. Nevertheless, spheres do not represent all local configurations of surfaces. Indeed, every smooth surface can be approximated at the second order with its so called osculating paraboloid of equation $z^2 = \lambda_1.x^2 + \lambda_2.y^2$ in an adequate Euclidean coordinate system. Its principal curvatures are λ_1 and λ_2. This is the reason why we have done some experiments on paraboloids of various principal curvatures that represent locally any smooth surfaces. We have chosen curvatures in the range of $[-0.1, 0.1]$ with an increment of 0.025. For each paraboloid, we have discretized it one hundred times, with an arbitrary rotation. The centre of the paraboloid was chosen randomly inside a voxel. For each family of discretized paraboloid, we present in Table 7.2 four values that are the minimum, maximum, average and ideal mean curvature computed at the surfel that is the closest from the origin of the paraboloid. First it should be noticed

Table 2. Experimental results for paraboloids. In each element of this array, from top to bottom, are given the minimum value, the maximum value, the average value of the mean curvature computed at the origin of the paraboloid of equation . The last value is the ideal mean curvature

λ_2 \ λ_1	-1.00e-01	-7.50e-02	-5.00e-02	-2.50e-02	0.00e+00	2.50e-02	5.00e-02	7.50e-02	1.00e-01
-1.00e-01	-1.26e-01 -1.15e-01 -1.21e-01 -1.00e-01	-1.13e-01 -1.05e-01 -1.10e-01 -8.75e-02	-1.03e-01 -9.26e-02 -9.81e-02 -7.50e-02	-8.79e-02 -7.65e-02 -8.25e-02 -6.25e-02	-7.02e-02 -5.35e-02 -6.30e-02 -5.00e-02	-4.90e-02 -3.29e-02 -4.13e-02 -3.75e-02	-2.95e-02 -1.35e-02 -2.40e-02 -2.50e-02	-1.75e-02 -1.73e-03 -1.02e-02 -1.25e-02	-9.28e-03 1.36e-02 1.68e-03 0
-7.50e-02		-1.05e-01 -9.49e-02 -9.99e-02 -7.50e-02	-9.13e-02 -8.42e-02 -8.76e-02 -6.25e-02	-7.68e-02 -6.64e-02 -7.19e-02 -5.00e-02	-5.71e-02 -4.52e-02 -5.11e-02 -3.75e-02	-3.54e-02 -2.22e-02 -3.01e-02 -2.50e-02	-1.89e-02 -3.19e-03 -1.24e-02 -1.25e-02	-6.46e-03 7.25e-03 1.18e-03 0	2.58e-03 2.03e-02 1.25e-02 1.25e-02
-5.00e-02			-7.84e-02 -7.19e-02 -7.51e-02 -5.00e-02	-6.22e-02 -5.59e-02 -5.89e-02 -3.75e-02	-4.12e-02 -3.22e-02 -3.79e-02 -2.50e-02	-1.95e-02 -1.02e-02 -1.61e-02 -1.25e-02	-5.39e-03 7.04e-03 1.54e-03 0	7.89e-03 2.21e-02 1.57e-02 1.25e-02	1.57e-02 3.50e-02 2.70e-02 2.50e-02
-2.50e-02				-4.45e-02 -4.05e-02 -4.23e-02 -2.50e-02	-2.28e-02 -1.57e-02 -2.02e-02 -1.25e-02	-2.47e-03 5.41e-03 1.77e-03 0	1.26e-02 2.37e-02 1.97e-02 1.25e-02	2.37e-02 3.93e-02 3.38e-02 2.50e-02	3.41e-02 5.39e-02 4.51e-02 3.75e-02
0.00e+00					-8.86e-04 1.02e-02 2.76e-03 0	1.99e-02 2.79e-02 2.43e-02 1.25e-02	3.39e-02 4.51e-02 4.12e-02 2.50e-02	4.68e-02 6.05e-02 5.47e-02 3.75e-02	5.51e-02 7.24e-02 6.55e-02 5.00e-02
2.50e-02						4.13e-02 4.70e-02 4.51e-02 2.50e-02	5.86e-02 6.40e-02 6.12e-02 3.75e-02	6.89e-02 7.83e-02 7.38e-02 5.00e-02	7.86e-02 9.01e-02 8.50e-02 6.25e-02
5.00e-02							7.36e-02 7.97e-02 7.72e-02 5.00e-02	8.49e-02 9.25e-02 8.96e-02 6.25e-02	9.57e-02 1.06e-01 9.99e-02 7.50e-02
7.50e-02								9.75e-02 1.06e-01 1.02e-01 7.50e-02	1.08e-01 1.16e-01 1.12e-01 8.75e-02
1.00e-01									1.18e-01 1.27e-01 1.22e-01 1.00e-01

that the results are more biased than for the sphere. They are overestimated in term of absolute value, but corresponds well to intuitive estimation of curvature that is less "local" than its mathematical definition (Figure 2(a) illustrates his point). Nevertheless, they still be quite insensitive to translation and rotation. The sign of the mean curvature is also well preserved.

8 Conclusion and perspectives

We have described in this paper an original and efficient technique that estimate the mean curvature field of a discrete surface. It depends on a scale parameter. Both time and space complexity is linear with respect to the number of surfels. Our method is fast enough to be useful in an interactive tool of manipulation of discrete surfaces, or in a multi-scale context. Our technique can be applied to object recognition and surface segmentation purposes. It may even be appropriate for measure when the surfaces curvature is regionally constant. Since the result strongly depends on the estimation of the slice contour curvature, further experiments with other methods of discrete plane contour curvature estimation should be done. The method of computation of differential characteristics may be adapted to work in higher dimensional spaces, to characterise the shape of discrete hypersurfaces since surface definition as well as slice contours still exist in these spaces. As an application, we have considered the surface of the human brain as a classical 2D grey level image where the grey level is the mean curvature and have extracted characteristic surface parts (Fig. 2(b) and Fig. 2(c)).

References

1. E. Artzy, G. Frieder, and G. T. Herman. The theory, design, implementation and evaluation of a three-dimensional surface detection algorithm. *CVGIP*, 15(6):1–24, 1981.
2. G. Bertrand and R. Malgouyres. Some topological properties of discrete surfaces. In *Proceedings of DGCI'96*, volume 1176 of *Lecture Notes in Computer Sciences*. Springer, 1996.
3. P. J. Besl and R.C. Jain. Invariant surface characteristic for 3d object recognition in range images. *CVGIP*, 33(999):33–80, 1986.
4. M. Do Carmo. *Differential geometry of curves and surfaces*. Prentice Hall, 1976.
5. R. Deriche. Recursively implementing the gaussian and its derivatives. Rapport de Recherche 1893, INRIA, April 1993.
6. T. J. Fan, G. Medioni, and R. Nevata. Recognising 3d objects using surface descriptions. *IEEE Transactions on Pattern Analysis and Machine Intelligence*, 11(11):1140–1157, November 1989.
7. G.T. Herman. Discrete multidimensional jordan surfaces. *CVGIP : Graphical Models and Image Processing*, 54(6):507–515, November 1992.
8. Y. Kerbrat and J.M. Braemer. *Géométrie des courbes et des surfaces*. Collection méthodes. Hermann, 1976.
9. A. Lenoir, R. Malgouyres, and M. Revenu. Fast computation of the normal vector field of the surface of a 3d discrete object. In *Proceedings of DGCI'96*, volume 1176 of *Lecture Notes in Computer Sciences*, pages 101–112. Springer, 1996.
10. P. Liang and C. H. Taubes. Orientation based differential geometric representations for computer vision applications. *IEEE Transactions on Pattern Analysis and Machine Intelligence*, 16(3), March 1994.
11. R. Malgouyres. A definition of surfaces in $\mathbb{Z}^3$. *Theorical Computer Sciences*. to appear.
12. O. Monga, N. Ayache, and P. T Sander. From voxel to curvature. Rapport de Recherche 1356, INRIA, December 1990.
13. O. Monga and S. Benayoun. Using partial derivatives of 3d images to extract typical surface features. *CVIU*, 61:171–189, 1995.
14. D.G. Morgenthaler and A. Rosenfeld. Surfaces in three-dimensional digital images. *Information and Control*, 51(3):227–247, 1981.
15. J. P. Thirion and A. Gourdon. Computing the differential characteristics of isointensity surfaces. *CVIU*, 61:190–202, 1995.
16. L. Thurfjellj, E. Bengtsson, and Bo Nordin. A boundary approach for fast neighbourhood operations on three- dimensional binary data. *CVGIP*, 57(1):13–19, 1995.
17. J. K. Udupa. Determination of 3d shape parameters from boundary information. *Computer Graphics and Image Processing*, 17:52–59, 1981.
18. J. K. Udupa. Multidimensional digital boundaries. *CVGIP: Graphical Models and Image Processing*, 56(4):311–323, July 1994.
19. J. K. Udupa and V .G. Ajjanagadde. Boundary and objet labelling in three-dimensional images. *CVGIP*, 51:355–369, 1990.
20. M. Worring and A.W.M. Smeulders. Digital curvature estimation. *CVGIP IU*, 58(3):366–382, 1993.
21. L. Yang, F. Albregtsen, and T. Taxt. Fast computation of three-dimensional geometric moments using a discrete divergence theorem and a generalisation to higher dimensions. *CVGIP*, 59(2):97–108, March 1997.

Ellipses Estimation from Their Digitization

Nataša Sladoje and Joviša Žunić
University of Novi Sad , Faculty of Engineering,
Trg D. Obradovića 6, 21000 Novi Sad, Yugoslavia
sladoje@uns.ns.ac.yu , ftn_zunic@uns.ns.ac.yu

Abstract

Ellipses in general position, and problems related to their reconstruction from digital data resulting from their digitization, are considered. If the ellipse

$$E : \tilde{A}(x-p)^2 + 2\tilde{B}(x-p)(y-q) + \tilde{C}(y-q)^2 \leq 1 , \quad \tilde{A}\tilde{C} - \tilde{B}^2 > 0 ,$$

is presented on digital picture of a given resolution, then the corresponding digital ellipse is:

$$D(E) = \left\{ (i,j) \mid A(i-a)^2 + 2B(i-a)(j-b) + C(j-b)^2 \leq r^2, \right.$$
$$\left. i,j \quad \text{are integers} \right\},$$

where r denotes the number of pixels per unit and $a = pr$, $b = qr$, $A = \tilde{A}r^2$, $B = \tilde{B}r^2$, $C = \tilde{C}r^2$.

Since the digitization of real shapes causes an inherent loss of information about the original objects, the precision of the original shape estimation from the corresponding digital data is limited, i.e. there is no possibility that the original ellipse can be recovered from the digital ellipse. What we present here is the estimation of parameters A, B, C and center position (a, b), of the ellipse digitized as above, with relative error bounded by $\mathcal{O}\left(\frac{1}{r^{15/11-\epsilon}}\right)$ and absolute error bounded by $\mathcal{O}\left(\frac{1}{r^{4/11-\epsilon}}\right)$, (where ϵ is an arbitrary positive number), that is, with the error tending to zero when the picture resolution increases. The obtained results imply that the half-axes of the original ellipse can be estimated with the same bounds of relative and absolute errors.

Index Terms – Pattern analysis, image processing, digital shapes, parameter estimation.

1   Introduction

Estimation of the relevant parameters of the original object, based on the data resulting from their digitization, is one of the the most important problems considered in computer vision and image processing ([2]). Digital shapes which appear the most often in practice are conic sections, including digital straight lines, (in the Euclidean plane) and so-called surfaces of the second order (in the Euclidean space). Some statistical parameter estimations of circular arcs can be founded in ([3]), bat the above mentioned problem, connected with the sets of digital points resulting from digitization of the ellipse in general position has not been solved by now. We find it might be interesting to give an efficient reconstruction of (general) ellipses by using digital data resulting from the digitization of the original object. Even though the problem of reconstructing corresponding digital ellipse still remains open, the method we describe provides obtaining the digital shape which can be considered as a good approximation of the coded one. For such purpose five properly chosen discrete moments are used.

The paper is organized as follows.

An asymptotical behaviour of discrete moments of digital ellipse, needed for the reconstruction and representation, is derived in the next section. An efficient estimation of the parameters of the original ellipse is given in Section 3. It is shown that the errors of such estimation tend to zero while r tends to infinity, i.e. when the resolution of digital picture increases. The confirming experimental results are presented: even for $r = 10$, relative deviation is shown to be less than 10% and for the absolute error we have 0.5 as the upper bound. For $r = 100$, relative error is less than 0.6%, and absolute error is bounded by 0.04.

2 Approximation of the Original Ellipse from its Digitization

Consider an ellipse E in the Euclidean plane, defined by the equation
$$\tilde{A}(x - p)^2 + 2\tilde{B}(x - p)(y - q) + \tilde{C}(y - q)^2 \le r^2 , \quad \tilde{A}\tilde{C} - \tilde{B}^2 > 0.$$
The ellipse E is digitized by using digitizing method in which all the digital points (points with integer coordinates) in the ellipse are taken. The data,resulting from the digitization of a given ellipse E obviously depend on the number of pixels per unit, i.e. on the picture resolution. Let r denotes this number; then the set of digital points, defined by

$$D(E) = \{(i,j) \mid A(i - a)^2 + 2B(i - a)(j - b) + C(j - b)^2 \le r^2,$$
$$i, j \text{ are integers } \} ,$$

is obtained as the digital picture of the ellipse E ($a = pr$, $b = qr$, $A = \tilde{A}r^2$, $B = \tilde{B}r^2$, $C = \tilde{C}r^2$). This set will be considered as the digital ellipse.

As different ellipses may have identical digital images, there is some indispensable uncertainty in retrieving the original ellipse from its digital image. So, a natural and important question is how efficiently the original ellipse E can be recovered from the data, resulting from its digitization $D(E)$? In this section it will be shown that certain information about digital ellipse $D(E)$ enables a reconstruction of a, b, A, B and C, with error tending to zero while $r \to \infty$.

A (k, l)-moment, denoted by $m_{k,l}(S)$ for a shape S, in 2D is defined by

$$m_{k,l}(S) = \iint\limits_{S} x^k y^l \, dx \, dy.$$

Considering that the parameters of the ellipse E can easily be reconstructed if $m_{0,0}(E)$, $m_{1,0}(E)$, $m_{0,1}(E)$ and any two of $m_{2,0}(E)$, $m_{1,1}(E)$ and $m_{0,2}(E)$ are known, and applying the analogous idea to the discrete shape, let us define the following integer parameters for a given ellipse E:

– the number of points of $D(E)$, denoted by $R(E)$;
– the sum of x-coordinates of points of $D(E)$, denoted by $X(E)$;
– the sum of y-coordinates of points of $D(E)$, denoted by $Y(E)$;
– the sum of squares of x-coordinates of points of $D(E)$, denoted by $XX(E)$;
– the sum of squares of y-coordinates of points of $D(E)$, denoted by $YY(E)$.

Obviously, these parameters can easily be computed and are uniquely determined for any ellipse. They can be understood as its discrete moments. So, it is natural to introduce the following definition:

Definition 1 *If a real ellipse E, given by $A(x-a)^2 + 2B(x-a)(y-b) + C(y-b)^2 \leq r^2$, is digitized, and $R(E)$, $X(E)$, $Y(E)$, $XX(E)$, $YY(E)$ are calculated from $D(E)$, then the estimated values, a_{est}, b_{est}, A_{est}, B_{est} and C_{est}, for the parameters of E are*

$$(a_{est}, b_{est}) = \left(\frac{X(E)}{R(E)}, \frac{Y(E)}{R(E)} \right);$$

$$A_{est} = \frac{4 \cdot r^2 \cdot \pi^2}{(R(E))^4} \cdot (YY(E) \cdot R(E) - (Y(E))^2);$$

$$B_{est} = \frac{4 \cdot r^2 \cdot \pi}{(R(E))^4} \cdot$$

$$\cdot \sqrt{16 \, \pi^2 \, (XX(E) \cdot R(E) - (X(E))^2) \cdot (YY(E) \cdot R(E) - (Y(E))^2) - (R(E))^6};$$

$$C_{est} = \frac{4 \cdot r^2 \cdot \pi^2}{(R(E))^4} \cdot (XX(E) \cdot R(E) - (X(E))^2).$$

It can be expected that the ellipse defined by

$$A_{est}(x - a_{est})^2 + 2B_{est}(x - a_{est})(y - b_{est}) + C_{est}(y - b_{est})^2 \leq r^2$$

is a good approximation of the original ellipse E. Its digitization, then, might be considered as a good approximation of $D(E)$.

In order to give the upper bound on the errors in estimating parameters of the original ellipse as defined above, we need asymptotical expressions for $R(E)$, $X(E)$, $Y(E)$, $XX(E)$ and $YY(E)$.

For convenience, in the rest of the paper it will be assumed that all the digital points have nonnegative coordinates. In other words, the origin is placed in the left-lower corner of the observed digital picture.

We cite the following result from the number theory ([1]).

Theorem 1 *If $\mathcal{B}$ is a convex body in R^2, with C^3 boundary and positive curvature at every point of the boundary, then the number of lattice points belonging to $r \cdot \mathcal{B}$ is:*

$$R(r \cdot \mathcal{B}) = r^2 \cdot P(\mathcal{B}) + \mathcal{O}\left(r^{\frac{7}{11}} \cdot (\log r)^{\frac{47}{22}} \right), \tag{1}$$

where $P(\mathcal{B})$ denotes the area of $\mathcal{B}$, while $r \cdot \mathcal{B}$ is dilatation of $\mathcal{B}$ by factor r.

What we will use is weaker: $R(r \cdot \mathcal{B}) = r^2 \cdot P(\mathcal{B}) + \mathcal{O}\left(r^{\frac{7}{11}+\epsilon}\right)$ for every $\epsilon > 0$. As a direct consequence is the asymptotical expression of $R(E)$:

$$R(E) \;=\; \sum_{\substack{i,j \text{ are integers} \\ A(i-a)^2 + 2B(i-a)(j-b) + C(j-b)^2 \le r^2}} 1 = \frac{\pi \cdot r^2}{\sqrt{AC - B^2}} + \mathcal{O}\left(r^{\frac{7}{11}+\epsilon}\right). \quad (2)$$

To give the asymptotical expressions for $X(E)$, $Y(E)$, $XX(E)$ and $YY(E)$, the following lemma will be needed.

Lemma 1 *Let an ellipse E be given by: $A(x-a)^2 + 2B(x-a)(y-b) + C(y-b)^2 \le r^2$, and $E_1(i)$ and $E_2(i)$ be determined by*

$E_1(i)$: $A(x-a)^2 + 2B(x-a)(y-b) + C(y-b)^2 \le r^2$ *and* $x \le i$,
where i is an integer   *and*
$E_2(i)$: $A(x-a)^2 + 2B(x-a)(y-b) + C(y-b)^2 \le r^2$ *and* $x \ge i$,
where i is an integer.

Let $R(E)$, $R(E_1(i))$ and $R(E_2(i))$ denote the numbers of digital points belonging to E, $E_1(i)$ and $E_2(i)$, respectively, while $P(E_1(i))$ and $P(E_2(i))$ denote the areas of E_1 and E_2, respectively. Then the following two (equivalent) relations

$$R(E_1(i)) \;=\; P(E_1(i)) + \frac{1}{2} \cdot L(E_1(i)) + \mathcal{O}\left(r^{\frac{7}{11}+\epsilon}\right) \quad and$$

$$R(E_2(i)) \;=\; P(E_2(i)) + \frac{1}{2} \cdot L(E_2(i)) + \mathcal{O}\left(r^{\frac{7}{11}+\epsilon}\right),$$

where

$$L(E_1(i)) = L(E_2(i)) = \frac{\sqrt{B^2(i-a)^2 - C(A(i-a)^2 - r^2)}}{C}$$

denotes the number of digital points of $D(E)$, lying on the line $x = i$, with the error bounded by 2, are satisfied.

Proof. The conditions of Theorem 1 can be relaxed to allow $\mathcal{B}$ having a finite number of corners, so (1) can be applied to the intersection of the interiors of two convex curves or, by subtraction, the region within one convex curve and outside another (for details, see [1]).

Let us consider the convex shape $\mathcal{L}(i)$, which is a subset of E, symmetrical with respect to the line $x = i$, and which contains all the digital points on the line $x = i$, within the ellipse E. Theorem 1 implies:

$$\frac{1}{2} \cdot (R(\mathcal{L}(i)) - L(E_1(i))) = \frac{1}{2} \cdot (P(\mathcal{L}(i)) - L(E_1(i))) + \mathcal{O}\left(r^{7/11+\epsilon}\right), \quad (3)$$

where $R(\mathcal{L}(i))$ denotes the number of digital points belonging to $\mathcal{L}(i)$, while $P(\mathcal{L}(i))$ denotes the area of $\mathcal{L}(i)$.

If $\bar{E}_1(i)$ denotes the (convex) shape $E_1(i) \cup \mathcal{L}(i)$, then, according to Theorem 1, for the number of digital points belonging to $\bar{E}_1(i)$, we have

$R(E_1(i)) + \frac{1}{2} \cdot (R(\mathcal{L}(i)) - L(E_1(i))) = P(E_1(i)) + \frac{1}{2} \cdot P(\mathcal{L}(i)) + \mathcal{O}\left(r^{7/11+\epsilon}\right)$.

Considering (3), the statement follows. $\square$

Now, the asymptotical expressions for $X(E)$ and $Y(E)$, can be given.

Theorem 2 *Let the real numbers a, b, A, B and C, satisfying $AC - B^2 > 0$, be given, and let all the values of x and y, satisfying $A(x-a)^2 + 2B(x-a)(y-b) + C(y-b)^2 \leq r^2$, be positive. Then the following asymptotical expressions hold:*

$$X(E) = \sum_{\substack{i,j \text{ are integers} \\ A(i-a)^2 + 2B(i-a)(j-b) + C(j-b)^2 \leq r^2}} i = \frac{a \cdot \pi \cdot r^2}{\sqrt{AC - B^2}} + \mathcal{O}\left(a \cdot r^{\frac{7}{11}+\epsilon}\right), (4)$$

$$Y(E) = \sum_{\substack{i,j \text{ are integers} \\ A(i-a)^2 + 2B(i-a)(j-b) + C(j-b)^2 \leq r^2}} j = \frac{b \cdot \pi \cdot r^2}{\sqrt{AC - B^2}} + \mathcal{O}\left(b \cdot r^{\frac{7}{11}+\epsilon}\right). (5)$$

Proof. Let's notice that $X(E)$ is equal to the number of digital points belonging to $3D$ body

$$\mathcal{C} = \{(x,y,z) \mid A(x-a)^2 + 2B(x-a)(y-b) + C(y-b)^2 \leq r^2, \quad 0 < z \leq x\} .$$

If $a - \dfrac{r\sqrt{C}}{\sqrt{AC - B^2}}$ is denoted by x_{min}, and $a + \dfrac{r\sqrt{C}}{\sqrt{AC - B^2}}$ is denoted by x_{max}, then for the points (x, y), satisfying $A(x-a)^2 + 2B(x-a)(y-b) + C(y-b)^2 \leq r^2$, we have $x \in [x_{min}, x_{max}]$.

Consider the number of digital points belonging to the body

$$\mathcal{C}' = \{(x,y,z) \mid A(x-a)^2 + 2B(x-a)(y-b) + C(y-b)^2 \leq r^2, x_{min} \leq z \leq x\} .$$

If $\mathcal{C}'$ is intersected by planes $z = \lceil x_{min} \rceil$, $z = \lceil x_{min} \rceil + 1$, ..., $z = \lfloor x_{max} \rfloor$, than, obviously, each digital point from $\mathcal{C}'$ belongs to one of those planes.

Expressing the volume of $\mathcal{C}'$ as the sum of the volumes obtained by previous intersectings, we get

$$\begin{aligned} vol(\mathcal{C}') &= \frac{a \cdot \pi \cdot r^2}{\sqrt{AC - B^2}} - x_{min} \cdot \frac{\pi \cdot r^2}{\sqrt{AC - B^2}} \\ &= \sum_{i=\lceil x_{min} \rceil}^{\lfloor x_{max} \rfloor - 1} vol(V_i) + (\lceil x_{min} \rceil - x_{min}) \cdot \frac{r^2 \cdot \pi}{\sqrt{AC - B^2}} + \mathcal{O}(r), \end{aligned} \tag{6}$$

where

$$\begin{aligned} V_i = \{(x,y,z) \mid &A(x-a)^2 + 2B(x-a)(y-b) + C(y-b)^2 \leq r^2, \\ &x \geq i, \quad i \leq z < \min\{x, i+1\}\}, \end{aligned}$$

for $i = \lceil x_{min} \rceil, \lceil x_{min} \rceil + 1, \ldots, \lfloor x_{max} \rfloor - 1$.

Since $vol(V_i) = P(E_2(i)) - vol(V_i')$,, where

$$\begin{aligned} V_i' = \{(x,y,z) \mid &A(x-a)^2 + 2B(x-a)(y-b) + C(y-b)^2 \leq r^2, \\ &x \geq i, \quad x \leq z < i+1\}, \end{aligned}$$

and, obviously,
$$\sum_{i=\lceil x_{min}\rceil}^{\lfloor x_{max}\rfloor-1} V_i' = \frac{1}{2}\cdot\frac{\pi\cdot r^2}{\sqrt{AC-B^2}}+\mathcal{O}(r),$$

by applying Lemma 1, we obtain

$$\sum_{i=\lceil x_{min}\rceil}^{\lfloor x_{max}\rfloor-1} vol(V_i) = \sum_{i=\lceil x_{min}\rceil}^{\lfloor x_{max}\rfloor-1}\left(R(E_2(i))-\frac{1}{2}L(E_2(i))+\mathcal{O}(r^{\frac{7}{11}+\epsilon})\right)$$

$$-\frac{1}{2}\cdot\frac{\pi\cdot r^2}{\sqrt{AC-B^2}}.$$

Further,

$$\sum_{i=\lceil x_{min}\rceil}^{\lfloor x_{max}\rfloor-1} L(E_2(i)) = \frac{\pi\cdot r^2}{\sqrt{AC-B^2}}+\mathcal{O}(r)\quad,$$

so (6) becomes

$$\frac{a\cdot\pi\cdot r^2}{\sqrt{AC-B^2}}-x_{min}\cdot\frac{\pi\cdot r^2}{\sqrt{AC-B^2}}=$$

$$=\sum_{i=\lceil x_{min}\rceil}^{\lfloor x_{max}\rfloor-1} R(E_2(i))-\frac{\pi\cdot r^2}{\sqrt{AC-B^2}}+(\lceil x_{min}\rceil-x_{min})\cdot\frac{r^2\cdot\pi}{\sqrt{AC-B^2}}+\mathcal{O}\left(r^{\frac{18}{11}+\epsilon}\right).$$

Finally, for the number of digital points in C', we have

$$\sum_{i=\lceil x_{min}\rceil}^{\lfloor x_{max}\rfloor-1} R(E_2(i)) = \frac{a\cdot\pi\cdot r^2}{\sqrt{AC-B^2}}-(\lceil x_{min}\rceil-1)\frac{r^2\cdot\pi}{\sqrt{AC-B^2}}+\mathcal{O}\left(r^{\frac{18}{11}+\epsilon}\right).\quad(7)$$

The number of digital points belonging to the body C'', determined by

$$A(x-a)^2+2B(x-a)(y-b)+C(y-b)^2\le r^2,\quad 0<z<x_{min}\quad,$$

is equal to

$$(\lceil x_{min}\rceil-1)\cdot\frac{\pi\cdot r^2}{\sqrt{AC-B^2}}+\mathcal{O}(a\cdot r^{\frac{7}{11}+\epsilon})$$

(since $x_{min}<a$), which, together with (7), completes the proof. The proof of the expression (5) is analogous. $\square$

Theorem 3 *Let the real numbers a, b, A, B and C, satisfying $AC-B^2>0$, be given, and let all the values of x and y, satisfying $A(x-a)^2+2B(x-a)(y-b)+$*

$C(y-b)^2 \leq r^2$, be positive. Then the following asymptotical expressions hold:

$$XX(E) \;=\; \sum_{\substack{i,j \text{ are integers} \\ A(i-a)^2+2B(i-a)(j-b)+C(j-b)^2 \leq r^2}} i^2$$

$$= \; \frac{\pi \cdot r^2}{\sqrt{AC-B^2}} \left(a^2 + \frac{C \cdot r^2}{4(AC-B^2)} \right) + \mathcal{O}(a^2 r^{\frac{7}{11}+\varepsilon}), \qquad (8)$$

$$YY(E) \;=\; \sum_{\substack{i,j \text{ are integers} \\ A(i-a)^2+2B(i-a)(j-b)+C(j-b)^2 \leq r^2}} j^2$$

$$= \; \frac{\pi \cdot r^2}{\sqrt{AC-B^2}} \left(b^2 + \frac{A \cdot r^2}{4(AC-B^2)} \right) + \mathcal{O}(b^2 r^{\frac{7}{11}+\varepsilon}). \qquad (9)$$

Proof. Let's notice that $XX(E)$ is equal to the number of digital points belonging to $3D$ body

$$\mathcal{G} = \left\{ (x,y,z) \mid A(x-a)^2 + 2B(x-a)(y-b) + C(y-b)^2 \leq r^2, \;\; 0 < z \leq x^2 \right\}.$$

As in the proof of Theorem 2, let $x \in [x_{min}, x_{max}]$ for the points (x,y) satisfying $A(x-a)^2 + 2B(x-a)(y-b) + C(y-b)^2 \leq r^2$. Consider the number of digital points belonging to the body $\mathcal{G}'$, given by

$$\mathcal{G}' \;:\; A(x-a)^2 + 2B(x-a)(y-b) + C(y-b)^2 \leq r^2, \;\; (x_{min})^2 < z \leq x^2.$$

If W_i, for $i = \lceil x_{min} \rceil, \lceil x_{min} \rceil + 1, \ldots, \lfloor x_{max} \rfloor - 1$, denotes the body

$$W_i \;=\; \left\{ (x,y,z) \mid A(x-a)^2 + 2B(x-a)(y-b) + C(y-b)^2 \leq r^2, \right.$$
$$\left. x \geq i, \;\; i^2 < z \leq \min\{x^2, (i+1)^2\} \right\},$$

then for the volume of $\mathcal{G}'$ we have

$$vol(\mathcal{G}') \;=\; \frac{\pi \cdot r^2}{\sqrt{AC-B^2}} \cdot \left(a^2 + \frac{C \cdot r^2}{4(AC-B^2)} \right) - (x_{min})^2 \cdot \frac{r^2 \cdot \pi}{\sqrt{AC-B^2}}$$

$$= \; \sum_{i=\lceil x_{min} \rceil}^{\lfloor x_{max} \rfloor - 1} vol(W_i) + ((\lceil x_{min} \rceil)^2 - (x_{min})^2) \cdot \frac{r^2 \cdot \pi}{\sqrt{AC-B^2}} + \mathcal{O}(a).$$

Since

$$vol(W_i) = ((i+1)^2 - i^2) \cdot P(E_2(i)) - vol(W_i') = (2i+1) \cdot P(E_2(i)) - vol(W_i')$$

where

$$W_i' \;=\; \left\{ (x,y,z) \mid A(x-a)^2 + 2B(x-a)(y-b) + C(y-b)^2 \leq r^2, \right.$$
$$\left. x \geq i, \;\; x^2 < z \leq (i+1)^2 \right\},$$

and

$$vol(W_i') = (2i+1) \cdot \frac{1}{2} \cdot L(E_2(i)) + (2i+1) \cdot \mathcal{O}\left(r^{\frac{1}{2}}\right) \ ,$$

we have, according to Lemma 1, that

$$\sum_{i=\lceil x_{min} \rceil}^{\lfloor x_{max} \rfloor - 1} vol(W_i) \quad = \quad \sum_{i=\lceil x_{min} \rceil}^{\lfloor x_{max} \rfloor - 1} (2i+1) \cdot (R(E_2(i)) - L(E_2(i)))$$

$$+ \sum_{i=\lceil x_{min} \rceil}^{\lfloor x_{max} \rfloor - 1} (2i+1) \cdot \mathcal{O}\left(r^{\frac{7}{11}+\epsilon}\right) \ .$$

So, the volume of $\mathcal{G}'$ can be expressed as

$$vol(\mathcal{G}') \quad = \quad \sum_{i=\lceil x_{min} \rceil}^{\lfloor x_{max} \rfloor - 1} (2i+1) \cdot (R(E_2(i)) - L(E_2(i)))$$

$$+ ((\lceil x_{min} \rceil)^2 - (x_{min})^2) \cdot \frac{\pi \cdot r^2}{\sqrt{AC - B^2}} + \mathcal{O}(a^2 \cdot r^{\frac{7}{11}+\epsilon})).$$

Notice that

$$\sum_{i=\lceil x_{min} \rceil}^{\lfloor x_{max} \rfloor - 1} (2i+1) \cdot (R(E_2(i)) - L(E_2(i))) + \mathcal{O}(a^2 \cdot r^{\frac{7}{11}+\epsilon}))$$

expresses the number of digital points in $\mathcal{G}'$, excluding the points belonging to the plane $z = (\lceil x_{min} \rceil)^2$.

Since the number of digital points belonging to

$$\mathcal{G}'' \quad = \quad \{(x,y,z) \mid A(x-a)^2 + 2B(x-a)(y-b) + C(y-b)^2 \le r^2,$$
$$0 < z \le (\lceil x_{min} \rceil)^2\}$$

is equal to $\quad (\lceil x_{min} \rceil)^2 \cdot \dfrac{\pi \cdot r^2}{\sqrt{AC - B^2}} + \mathcal{O}(a^2 \cdot r^{\frac{7}{11}+\epsilon}) \quad$, the expression for the number of digital points in $\mathcal{G} = \mathcal{G}' \cup \mathcal{G}''$ is

$$XX(E) \quad = \quad \sum_{i=\lceil x_{min} \rceil}^{\lfloor x_{max} \rfloor - 1} (2i+1) \cdot (R(E_2(i)) - L(E_2(i))) + (\lceil x_{min} \rceil)^2 \cdot \frac{\pi \cdot r^2}{\sqrt{AC - B^2}}$$
$$+ \mathcal{O}(a^2 \cdot r^{\frac{7}{11}+\epsilon}).$$

Then, by using expression for $vol(\mathcal{G}')$, we have

$$XX(E) = \frac{\pi \cdot r^2}{\sqrt{AC - B^2}} \cdot \left(a^2 + \frac{C \cdot r^2}{4(AC - B^2)}\right) + \mathcal{O}(a^2 \cdot r^{\frac{7}{11}+\epsilon}) \ ,$$

which completes the proof. The proof of the expression (9) is analogous. $\quad\square$

3 Errors in Estimating the Original Ellipse E from its Discrete Moments $R(E)$, $X(E)$, $Y(E)$, $XX(E)$ and $YY(E)$

Possible difference in the order of magnitude of a, b and r, which is not excluded by the assumptions of Theorem 2, Theorem 3 and Theorem 4, affects the order of the error in estimating the parameters of the original ellipse as it is defined in Definition 1, if the asymptotical expressions for discrete moments $R(E)$, $X(E)$, $Y(E)$, $XX(E)$ and $YY(E)$ are directly substituted. However, as a consequence of the following statement, it can be assumed that a, b and r are of the same order.

Theorem 4 Let the ellipse $E = E(a, b, A, B, C) : A(x - a)^2 + 2B(x - a)(y - b) + C(y - b)^2 \leq r^2$ be given. Then the the the values $XX(E) \cdot R(E) - (X(E))^2$, $YY(E) \cdot R(E) - (Y(E))^2$, $R(E)$, $\frac{X(E)}{R(E)} - a$ and $\frac{Y(E)}{R(E)} - b$ are the constants with respect to a translation by a vector having integer components. In other words, the following equalities are satisfied:

a) $XX(E(a, b, A, B, C)) \cdot R(E(a, b, A, B, C)) - (X(E(a, b, A, B, C)))^2 =$
$= XX(E(a + k, b + l, A, B, C)) \cdot R(E(a + k, b + l, A, B, C))$
$- (X(E(a + k, b + l, A, B, C)))^2$,

b) $YY(E(a, b, A, B, C)) \cdot R(E(a, b, A, B, C)) - (Y(E(a, b, A, B, C)))^2 =$
$= YY(E(a + k, b + l, A, B, C)) \cdot R(E(a + k, b + l, A, B, C))$
$- (Y(E(a + k, b + l, A, B, C)))^2$,

c) $R(E(a, b, A, B, C)) = R(E(a + k, b + l, A, B, C))$,

d) $\frac{X(E(a,b,A,B,C))}{R(E(a,b,A,B,C))} - a = \frac{X(E(a+k,b+l,A,B,C))}{R(E(a+k,b+l,A,B,C))} - (a + k)$,

e) $\frac{Y(E(a,b,A,B,C))}{R(E(a,b,A,B,C))} - b = \frac{Y(E(a+k,b+l,A,B,C))}{R(E(a+k,b+l,A,B,C))} - (b + l)$,

where k and l are integers.

Proof. a) For $k = 1$ and $l = 0$, the relation follows by using equalities

$$\sum_{(i,j) \in D(E(a+1,b,A,B,C))} 1 = \sum_{(i,j) \in D(E(a,b,A,B,C))} 1 \quad ;$$

$$\sum_{(i,j) \in D(E(a+1,b,Ar,Br))} i = \sum_{(i,j) \in D(E(a,b,A,B,C))} (i + 1) \quad ,$$

while other cases follow by induction or as a consequence of symmetry.

The proof of b) is analogous; the proofs of c), d) and e) are trivial. $\square$

Now, we give an upper bound on the precision of estimation the parameters of the ellipse from five discrete moments, $R(E)$, $X(E)$, $Y(E)$, $XX(E)$ and $YY(E)$, corresponding to its digitization.

Theorem 5 *In estimating parameters of the ellipse E, given by* $A(x - a)^2 + 2B(x - a)(y - b) + C(y - b)^2 \leq r^2$ *from its digitization* $D(E)$, *the relative errors*

$$\frac{a_{est}}{a} - 1 \ , \quad \frac{b_{est}}{b} - 1 \ , \quad \frac{A_{est}}{A} - 1 \ , \quad \frac{B_{est}}{B} - 1 \quad and \quad \frac{C_{est}}{C} - 1$$

are upper bounded by $\mathcal{O}\left(\dfrac{1}{r^{\frac{15}{11} - \epsilon}}\right)$.

Proof. a) and b) follow directly from Definition 1 and (2)-(4). Since the same order of magnitude for a, b and r can be assumed, so c), d) and e) follow from Definition 1, (2)-(4), (7) and (8). ☐

Numerical results are presented in Table 1.

It is not surprising that the relative errors tend to zero while r tends to infinity, but convergence of the absolute errors to zero, with r tending to infinity, might be unexpected. Anyway, the following results show that the absolute errors in estimating parameters of the digitized ellipse, just as the relative errors, tend to zero when the resolution of the digital picture is high.

Theorem 6 *If the ellipse* $E : A(x - a)^2 + 2B(x - a)(y - b) + C(y - b)^2 \leq r^2$ *is digitized and its parameters are estimated from* $D(E)$, *then for the absolute errors we have*

$$a_{est} - a \quad and \quad b_{est} - b \quad are \ upper \ bounded \ by \quad \mathcal{O}\left(\tfrac{1}{r^{\frac{4}{11} - \epsilon}}\right), \quad while$$

$$A_{est} - A \ , \quad B_{est} - B \quad and \quad C_{est} - C \quad are \ upper \ bounded \ by \quad \mathcal{O}\left(\tfrac{1}{r^{\frac{15}{11} - \epsilon}}\right).$$

Proof. By using Definition 1, the expressions for the appearing discrete moments and considering the assumption that r, a and b have the same order of magnitude, the statement follows. ☐

Table 2. contains the experimental results.

Since the slight change of the parameters a, b, A, B and C implies the slight change of the half-axes and the angle of rotation transforming given ellipse to the position where its axes are parallel to coordinate axes, which means that the slight change of a, b, A, B and C leads to the ellipse with slightly changed form and position, the digitization of the reconstructed ellipse can be considered as a good approximation of the coded one.

Moreover, we mention that the previous results imply that the half-axes of the original ellipse can be estimated with the same bound of the relative error as it is obtained for the center position, (a, b), and parameters A, B and C. Namely, if the half-axes of the original ellipse are denoted by $\tilde{A}$ and $\tilde{B}$, then the relations $\frac{1}{\tilde{A}^2} + \frac{1}{\tilde{B}^2} = A + C$ and $\frac{1}{\tilde{A}^2} \cdot \frac{1}{\tilde{B}^2} = AC - B^2$ are satisfied, so $\frac{1}{\tilde{A}^2}$ and $\frac{1}{\tilde{B}^2}$ can be obtained as the solution of the equation $x^2 - (A + C)x + AC - B^2 = 0$. It is easy to conclude, then, that $\tilde{A}$ and $\tilde{B}$ can be recovered (by using A_{est}, B_{est}

and C_{est}) with the $\mathcal{O}\left(\frac{1}{r^{\frac{11}{11}-\epsilon}}\right)$ relative error. Similarly, it can be derived that the absolute errors of such estimation of the half-axes are bounded by $\mathcal{O}\left(\frac{1}{r^{\frac{11}{11}-\epsilon}}\right)$.

Relative error in estimating the parameters A, B, C and (a, b) of the ellipse
$$E: A(x-a)^2 + 2B(x-a)(y-b) + C(y-b)^2 \leq r^2,$$
by A_{est}, B_{est}, C_{est} and (a_{est}, b_{est}), respectively;
$$MAX_{rel\_err} = \max\{\left|\frac{A_{est}}{A}-1\right|, \left|\frac{B_{est}}{B}-1\right|, \left|\frac{C_{est}}{C}-1\right|, \left|\frac{a_{est}}{a}-1\right|, \left|\frac{b_{est}}{b}-1\right|\}$$

A	B	C	a	b	r	MAX_{rel_err}
					10	0.08916
7.5	5.2	11.7	57.7	76.6	100	0.00297
					1000	0.00032
					10	0.05817
4.5	1.3	5.7	1615.9	849.6	100	0.00519
					1000	0.00011
					10	0.00602
0.9	0.3	0.7	615.9	89.6	100	0.00045
					1000	0.00001

Table 1.

Absolute error in estimating the parameters A, B, C and (a, b) of the ellipse
$$E: A(x-a)^2 + 2B(x-a)(y-b) + C(y-b)^2 \leq r^2,$$
by A_{est}, B_{est}, C_{est} and (a_{est}, b_{est}), respectively;
$$MAX_{abs\_err} = \max\{|A_{est}-A|, |B_{est}-B|, |C_{est}-C|, |a_{est}-a|, |b_{est}-b|\}$$

A	B	C	a	b	r	MAX_{abs_err}
					10	0.49874
7.5	5.2	11.7	57.7	76.6	100	0.02207
					1000	0.01868
					10	0.21307
4.5	1.3	5.7	1615.9	849.6	100	0.03291
					1000	0.01272
					10	0.03488
0.9	0.3	0.7	615.9	89.6	100	0.02942
					1000	0.00311

Table 2.

4 Comments and Conclusion

Since the ellipses appear very often in practice of computer vision and image processing, the solution of the problem of their efficient reconstruction from the corresponding digital picture might be useful. Mentioning that the digitization of real shapes causes an inherent loss of information about the original objects, which implies that the precision of the estimation of the original shape from the corresponding digital data is limited, in previous sections we presented a constant time reconstruction of the parameters in the equation of the original ellipse with $\mathcal{O}\left(\frac{1}{r^{\frac{15}{11}-\varepsilon}}\right)$-relative error and $\mathcal{O}\left(\frac{1}{r^{\frac{4}{11}-\varepsilon}}\right)$-absolute error, where ε is an arbitrary positive number, and r is the number of pixels per unit, which means that the errors of such reconstruction tend to zero when the the picture resolution is high. Consequently, the half-axes of the original ellipse can be reconstructed from the estimated values of the parameters A, B and C, with the errors (both, relative and absolute) of the same order as the parameters, themselves. As a result, the coded digital ellipse can be approximated by the digitization of the reconstructed ellipse. Five discrete moments, corresponding to the digitization of the original ellipse, are needed for that purpose.

References

[1] M. N. Huxley, "Exponential Sums and Lattice Points", *Proc. London Math. Soc.* (3), 60, 471–502, 1990.

[2] J. Koplowitz and A. Bruckstein "Design of Perimeter Estimators for Digitized Planar Shapes", *IEEE Trans. Pattern Analysis and Machine Intelligence*, vol. 11, pp. 611-622, 1989.

[3] M. Worring and A.W.M. Smeulders, " Digitized circular arcs: characterization and parameter estimation", *IEEE Trans. PAMI*, vol. 17, pp. 587-598, 1995.

Topological Errors and Optimal Chamfer Distance Coefficients[*]

Stéphane Marchand-Maillet and Yazid M. Sharaiha

Operational Research and Systems - Imperial College
53 Prince's Gate - London SW7 2PG - United Kingdom
s.m.marchand@ic.ac.uk – http://www.ms.ic.ac.uk/steph/

Abstract. In this paper, a theoretical characterisation of the topological errors which arise during the approximation of Euclidean distances from discrete ones is presented. The continuous distance considered is the widely used Euclidean distance whereas we consider as discrete distance the chamfer distance based on 3×3 masks. The objective is to obtain formal results from which algorithms for the exact solution of the Euclidean Distance Transformation using integer arithmetic can be derived. We conclude this study by presenting a global upper bound for a topologically-correct distance mapping, irrespective of the chamfer distance coefficients, and identify the smallest coefficients associated with this bound.

1   Introduction

The main motivation of this work is to analyse the mapping between continuous (Euclidean) and discrete (chamfer) distances on the unit square grid [7].

Section 2 first recalls general digital image processing terminology and the principles behind approximating continuous distances by discrete ones. In [3,4], empirical results were presented to point out that the pixel ordering induced by the chamfer (discrete) Distance Transformation (DT) only matches with the ordering induced by the Euclidean (continuous) Distance Transformation up to some upper bound. In Section 3, we establish this in mathematical terms and obtain a closed form solution for such upper bounds for any given DT coefficients. The objective is first to characterise the topological errors in the mapping between continuous and discrete distances, and then to derive distance bounds which guarantee an error-free transformation. Finally, Section 3.3 details the characterisation of integer DT coefficients which are proved to be optimal for such a criterion.

While studying in depth the calculation of Euclidean distance values using discrete distance functions, we will derive results concerning the decomposition of integer values which can then form the basis for the development of exact Euclidean Distance Transformation algorithms (see *e.g.*, [8]).

---

[*] Work supported by the EPSRC, UK (grant reference number GR/J85271).

2 Definitions and Notations

We consider throughout these pages that the continuous distance used is the Euclidean distance d_E defined as $d_E(p,q) = \sqrt{(x_p - x_q)^2 + (y_p - y_q)^2}$, where $p = (x_p, y_p)$ and $q = (x_q, y_q)$. We introduce the notation for some standard functions. $\lceil x \rceil$ is the smallest integer greater or equal to than $x \in \mathbb{R}$ and $\lfloor x \rfloor$ is greatest integer smaller or equal to than $x \in \mathbb{R}$. Then, round$(x) = \lfloor x \rfloor$ if $\lfloor x \rfloor \leq x \leq \lfloor x \rfloor + \frac{1}{2}$. Otherwise, round$(x) = \lceil x \rceil$.

In approximating the Euclidean distance on the discrete grid, chamfer distances were introduced in [5,6] and studied in [1,2]. Such discrete distances typically rely on the definition of local distances within a mask centred at each pixel. We will consider chamfer distances in relation to 3×3 masks. We define a as the length of the unit horizontal/vertical move (a-move) on the grid, and b as the length of the unit diagonal move (b-move) on the grid. Given two points p and q, the chamfer distance $d_{a,b}(p,q)$ between p and q can be computed as follows.

$$d_{a,b}(p,q) = k_a a + k_b b \tag{1}$$

where k_a and k_b represent the number of a- and b-moves on the shortest path from p to q on the grid. The conditions on a and b for $d_{a,b}$ to be a distance are given in (2) below (see [9] and [12] for more details).

$$0 < a < b < 2a \tag{2}$$

The number of a- and b-moves (k_a, k_b) on the shortest path from p to q can also be used to compute the Euclidean distance between p and q.

$$d_E(p,q) = \sqrt{(k_a + k_b)^2 + k_b^2} \tag{3}$$

Without loss of generality, we restrict this study to a and b values such that a and b are relatively prime (*i.e.*, the Greater Common Divisor of a and b, gcd(a,b), is such that gcd$(a,b) = 1$). This corresponds to normalising the a and b coefficients to their minimal configuration.

Forchhammer [3,4] pointed out that topological errors occurred when deducing a Euclidean Distance Map from a Discrete Distance Map. This work was based on the distance inequalities to be satisfied during the generation of the Distance Map. From this study, he derived empirical results concerning the limitations of discrete distances in approximating continuous ones. In the next section, we formally detail these topological errors induced by the approximation of continuous distances by discrete ones and present distance limits as upper bounds for the correctness of the Distance Maps.

3 Topological Errors

In [3,4], Forchhammer introduced the topological inconsistencies induced by the discrete distances when used as an approximation of the Euclidean distance.

Essentially, the ordering of the discrete distance does not match the ordering of the Euclidean distance. Consider the following example (see Fig. 1). Let the DT coefficients be $a = 2$, $b = 3$, and consider the three integer points (pixels), $p = (0,0)$, $q = (10,1)$ and $r = (9,4)$. The shortest path on the grid from p to q is given by $k_a = 9$ and $k_b = 1$ and that from p to r is given by $k_a = 5$ and $k_b = 4$. Using Equations (1) and (3), we have $d_{a,b}(p,q) = 21$, $d_E(p,q) = \sqrt{101}$, $d_{a,b}(p,r) = 22$ and $d_E(p,r) = \sqrt{97}$. If q and r are border pixels, the discrete DT will lead to consider q as the nearest border pixel to p (by the chamfer distance measure) giving an approximate Euclidean distance of $\sqrt{101}$. This is clearly incorrect since there is a smaller Euclidean distance between p and another border pixel (namely r) giving a Euclidean distance of $\sqrt{97}$. In other words, since, $d_{a,b}(p,q) < d_{a,b}(p,r)$ and $d_E(p,q) > d_E(p,r)$, the ordering of $d_{a,b}$ differs from the ordering of d_E.

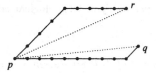

Fig. 1. An example of a topological error.

Given a pair of DT coefficients (a, b), we characterise the configurations for which this problem occurs precisely. First, we introduce how restrictions for the decomposition of a given discrete distance value D into a- and b-moves can be given by the solution to the Frobenius problem (see [11]).

Theorem 1. *[11]. Given $0 < a < b$ such that $(a, b) \in \mathbb{N}^2$ and $\gcd(a, b) = 1$. Consider the equation: $k_a a + k_b b = D$ $(k_a, k_b) \in \mathbb{N}^2$ If $\chi = (a - 1)(b - 1)$, then we have the following instances.*

(i) If $D \geq \chi$, there is always at least one solution (k_a, k_b).
(ii) If $D = \chi - 1$, there is no solution.
(iii) There is exactly $\frac{1}{2}\chi$ values of D that have no solution.

We will use the solution to this classical problem to characterise the topological error introduced earlier. Two types of errors are distinguished and presented in Sections 3.1 and 3.2 respectively.

3.1 Type 1 error

Given a pair of DT coefficients (a, b) and three integer points p, q and r, a Type 1 error occurs between q and r relative to p if $d_{a,b}(p,q) = d_{a,b}(p,r)$ and $d_E(p,q) \neq d_E(p,r)$. More formally, we make the following definition.

Definition 2. Type 1 error. Given a pair of DT coefficients (a, b) and a discrete distance value D, a Type 1 topological error occurs if there exist two integer pairs (k_{a_1}, k_{b_1}) and (k_{a_2}, k_{b_2}) such that $k_{a_1}a + k_{b_1}b = k_{a_2}a + k_{b_2}b = D$ and $\sqrt{(k_{a_1} + k_{b_1})^2 + k_{b_1}^2} \neq \sqrt{(k_{a_2} + k_{b_2})^2 + k_{b_2}^2}$.

An example for Type 1 error is illustrated in Fig. 2 where, (k_{a_1}, k_{b_1}) represents the shortest path from p to q, and (k_{a_2}, k_{b_2}) the shortest path from p to r. In this example, the DT coefficients are $a = 2$ and $b = 3$, and the three integer points are $p = (0,0)$, $q = (3,0)$ and $r = (2,2)$. We obtain $d_{a,b}(p,q) = d_{a,b}(p,r) = D = 6$, since $k_{a_1} = 3$, $k_{b_1} = 0$ and $k_{a_2} = 3$, $k_{b_2} = 0$. On the other hand, we have, $d_E(p,q) = 3$ and $d_E(p,r) = \sqrt{8}$.

Fig. 2. The first instance of Type 1 topological error for $(a,b) = (2,3)$.

Lemma 3. [7] *Given a pair of DT coefficients (a,b) and a discrete distance value D, we assume that the existence condition (i) in Theorem 1 holds. Then, the maximal value $k_{b_{max}}$ of k_b such that $k_a a + k_{b_{max}} b = D$ with $k_a \geq 0$ and $k_{b_{max}} \geq 0$ is given by:*

$$k_{b_{max}} = \left\lfloor \frac{D}{b} \right\rfloor - \psi((D \bmod b) \bmod a) \qquad (4)$$

where ψ is the implicit integer function such that:
$$\psi : \{0, 1, \cdots, a-1\} \mapsto \{0, 1, \cdots, a-1\} \text{ and } \psi((x.(2a-b)) \bmod a) = x.$$

ψ can be easily calculated as a one-to-one mapping of the set $\{0, 1, \cdots, a-1\}$ onto itself. For example (see Fig. 3), if $a = 5$ and $b = 7$, the mapping is given by $\{0,1,2,3,4\} \overset{\psi}{\mapsto} \{0,2,4,1,3\}$. Therefore, if $D = 86$, say, we have $\lfloor \frac{D}{b} \rfloor = 12$ and $(D \bmod b) \bmod a = 2$. Hence, from Equation (4), $k_{b_{max}} = 12 - \psi(2) = 12 - 4 = 8$. We can also easily compute $k_a = \frac{D - k_{b_{max}} b}{a} = 6$. Lemma 3 would allow us to have all the values of k_b since these can be given by $(k_{b_{max}} - ia)$ with $i = 0, \cdots, \lfloor \frac{k_{b_{max}}}{a} \rfloor$. Therefore, the pairs (k_a, k_b) for the decomposition of $D = 86$ are (6,8) and (13,3).

Thus, Lemma 3 readily gives an exhaustive list of (k_a, k_b) pairs for decomposing any discrete distance value D for any DT coefficients a and b. Note that, if Condition *(i)* in Theorem 1 is not matched (*i.e.*, no possible decomposition), we obtain $k_{b_{max}} < 0$ (*e.g.*, $a = 3$, $b = 4$, $\chi = 6$, if $D = 5 = \chi - 1$, we obtain $k_{b_{max}} = -1$). This simple test can prove useful when developing the mapping algorithms.

Definition 4. Given a pair of DT coefficients (a,b) and a discrete distance value D which can be decomposed in at least one manner, we define:

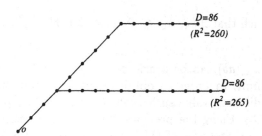

Fig. 3. All decompositions for $D = 86$.

- The set $\Theta = \{(k_{a_i}, k_{b_i}), i = 0, \cdots, n\}$ as the exhaustive list of all possible decomposition pairs (*i.e.*, $D = k_{a_i}a + k_{b_i}b$, $k_{a_i} \geq 0$, $k_{b_i} \geq 0 \ \forall i = 0, \cdots, n$ with $n = \left\lfloor \frac{k_{b_{max}}}{a} \right\rfloor$). Note that $k_{b_{max}} = \max_{i=0,\cdots,n} k_{b_i}$ and, $k_{b_i} = k_{b_{max}} - ia$. Therefore, the set Θ can be fully computed using Lemma 3.
- $R_i(D)$ as the Euclidean distance associated with the pair (k_{a_i}, k_{b_i}). From Equation (3), we have,

$$R_i(D) = \sqrt{(k_{a_i} + k_{b_i})^2 + k_{b_i}^2} \tag{5}$$

- $R_{max}(D)$ (*resp.* $R_{min}(D)$) as the maximal (*resp.* minimal) Euclidean distance over all $n + 1$ possible decompositions.
- l (*resp.* m) as the index in the set Θ of the decomposition (k_{a_l}, k_{b_l}) (*resp.* (k_{a_m}, k_{b_m})) leading to $R_{max}(D)$ (*resp.* $R_{min}(D)$).

Continuing with the example in Fig. 3, we had $a = 5$, $b = 7$ and $D = 86$. We obtained $k_{b_{max}} = 8$, $n = 1$ and $\Theta = \{(6,8),(13,3)\}$ (*i.e.*, $R_0(86) = \sqrt{260}$ and $R_1(86) = \sqrt{265}$). Hence, $m = 0$, $l = 1$ (*i.e.*, $R_{min}(D) = R_0(86) = \sqrt{260}$ and $R_{max}(D) = R_1(86) = \sqrt{265}$). Note that in [7], formulae for calculating m and l without the need of enumeration were derived.

Lemma 5. *Characterisation of a Type 1 error. Given a pair of DT coefficients (a, b), a Type 1 error occurs for any discrete distance value D for which $R_{max}(D) \neq R_{min}(D)$.*

Clearly, the first instance of D for which $R_{max}(D) \neq R_{min}(D)$ is $D = ab$. In this case, $k_{b_{max}} = a$, $n = 1$, $\Theta = \{(0, a), (b, 0)\}$, (*i.e.*, $R_0 = a\sqrt{2}$ and $R_1 = b$). Hence, $R_{max}(ab) \neq R_{min}(ab)$. Therefore, we define the following Euclidean distance limit when considering Type 1 errors only.

Definition 6. Euclidean distance limit induced by Type 1 error, $\mathcal{R}_1(a, b)$. Given a pair of DT coefficients (a, b), and $D = ab$ as the minimum discrete distance value for which $R_{min}(D) \neq R_{max}(D)$, we define the Euclidean distance limit $\mathcal{R}_1(a, b)$ for Type 1 errors as follows.

$\mathcal{R}_1(a, b)$ is the maximal Euclidean distance value deduced from a discrete distance value (*i.e.*, using Equation (3)) up to which both discrete and continuous distance ordering match. More formally, $\mathcal{R}_1(a, b)$ is the maximal Euclidean distance value R such that $\exists\, k_a,\ k_b \in \mathbb{N}$ such that $R = \sqrt{(k_a + k_b)^2 + k_b^2}$ and $R < R_{\max}(ab)$.

In other words, $R_{\max}(ab)$ can be considered as a strict (*i.e.*, non-feasible) Euclidean distance limit. In order to obtain a feasible limit, we search D' the maximal discrete distance value such that $R_{\min}(D') < R_{\max}(ab)$ and consider $\mathcal{R}_1(a, b) = R_{\min}(D')$. Using the previous study, we can easily design an algorithm to compute, for any pair of DT coefficients (a, b), the value of $\mathcal{R}_1(a, b)$. In Fig. 4, $(\mathcal{R}_1(a, b))^2$ is plotted for each pair of valid DT coefficients such that $a \le 10$.

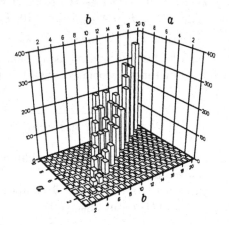

Fig. 4. Euclidean distance limit induced by Type 1 of topological error $(\mathcal{R}_1(a, b))^2$.

The following table gives the corresponding discrete distance values $D = ab$ and D' for some instances of DT coefficients (a, b).

a	b	$D = ab$	D'	$R_{\max}(ab)$	$\mathcal{R}_1(a, b) = R_{\min}(D')$
2	3	6	6	$\sqrt{9}$	$\sqrt{8}$
3	4	12	13	$\sqrt{18}$	$\sqrt{17}$
3	5	15	16	$\sqrt{25}$	$\sqrt{20}$
5	7	35	36	$\sqrt{50}$	$\sqrt{45}$

3.2 Type 2 error

Given a pair of DT coefficients (a, b) and three integer points p, q and r, a Type 2 error occurs between q and r, relative to p, if $d_{a,b}(p, q) < d_{a,b}(p, r)$ and $d_{\mathrm{E}}(p, q) > d_{\mathrm{E}}(p, r)$. More formally, we make the following definition.

Definition 7. Type 2 error. Given a pair of DT coefficients (a, b) and two discrete distance values D_1 and D_2, we assume that $\exists\, (k_{a_1}, k_{b_1})$ and (k_{a_2}, k_{b_2}) such that $k_{a_1}a + k_{b_1}b = D_1$ and $k_{a_2}a + k_{b_2}b = D_2$ (see Theorem 1). A Type 2 error occurs if, $D_1 < D_2$ and $\sqrt{(k_{a_1}+k_{b_1})^2 + k_{b_1}^2} > \sqrt{(k_{a_2}+k_{b_2})^2 + k_{b_2}^2}$.

Fig. 5 illustrates an instance of Type 2 error between q and r, relative to p. In this example, the DT coefficients are $a = 2$ and $b = 3$. We consider $p = (0,0)$, $q = (6,0)$ and $r = (5,3)$. Therefore, $D_1 = d_{a,b}(p,q) = 12$, since $k_{a_1} = 6$ and $k_{b_1} = 0$. We also obtain $D_2 = d_{a,b}(p,r) = 13$, since $k_{a_2} = 2$ and $k_{b_2} = 3$. On the other hand, we have $d_E(p,q) = \sqrt{36}$ and $d_E(p,r) = \sqrt{34}$. Therefore, $d_{a,b}(p,q) < d_{a,b}(p,r)$ and $d_E(p,q) > d_E(p,r)$.

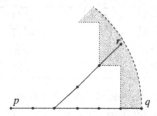

Fig. 5. The first instance of Type 2 topological error for $(a,b) = (2,3)$.

From a geometric viewpoint, given three integer points p, q and r such that $D_1 = d_{a,b}(p,q) < D_2 = d_{a,b}(p,r)$, a Type 2 error occurs between q and r, relative to p, if there exists at least one integer point (namely, r) included in the area between the discrete disc of radius D_1 centred at p and the Euclidean disc that contains this discrete disc. In Fig. 5, this area is illustrated by the shaded surface. A Type 2 error occurs between q and r, relative to p, since r lies in this shaded surface.

The geometrical characterisation of Type 2 errors will, therefore, be investigated through the characterisation of the radius of the smallest Euclidean disc that contains the discrete disc of radius D for any given values of the DT coefficients (a, b) and for any discrete distance value D. The radius of such a Euclidean disc was noted $R_{max}(D)$ in Section 3.1 (see Definition 4). Therefore, the geometrical characterisation of Type 2 error can be formally written as follows.

Lemma 8. *Given a pair of DT coefficients (a, b) and a discrete distance value D, a Type 2 error occurs in the discrete disc of radius D if there exists a discrete distance value $D' > D$ such that $R_{min}(D') < R_{max}(D)$.*

Using this characterisation, the Euclidean distance limit $\mathcal{R}_2(a, b)$ induced by Type 2 error can be defined as follows.

Definition 9. Euclidean distance limit induced by Type 2 error, $\mathcal{R}_2(a, b)$ Given a pair of DT coefficients (a, b), and D, the minimum discrete distance value for which there exists a discrete distance value D' such that $R_{min}(D') < R_{max}(D)$, we define the Euclidean distance limit $\mathcal{R}_2(a, b)$ for Type 2 errors as follows.

$\mathcal{R}_2(a, b) = R_{min}(D')$ where D' is the smallest discrete distance value such that $R_{min}(D') < R_{max}(D)$.

In other words, if a Type 2 error occurs for the discrete distance value D (e.g., at point q in Fig. 5, with $D = 12$), we consider the Euclidean distance limit as the value $R_{\min}(D')$ where D' is the discrete distance value at the second point for which Type 2 error occurred (e.g., point r in Fig. 5, and $D' = 13$).

Using the previous study, we can also design an algorithm to compute, for any pair of DT coefficients (a, b), the value of $\mathcal{R}_2(a, b)$. In Fig. 6(A), $(\mathcal{R}_2(a, b))^2$ is plotted for each DT coefficients pair such that $a \leq 10$. The table below gives the corresponding discrete distance values D and D' for some instances of DT coefficients (a, b).

a	b	D	D'	$R_{\max}(D)$	$\mathcal{R}_2(a, b) = R_{\min}(D')$
2	3	12	13	$\sqrt{36}$	$\sqrt{34}$
3	4	12	13	$\sqrt{18}$	$\sqrt{17}$
3	5	9	10	$\sqrt{9}$	$\sqrt{8}$
5	7	21	22	$\sqrt{18}$	$\sqrt{17}$

Using the results of Sections 3.1 and 3.2, we can now define a combined Euclidean distance limit where no topological error of any type can occur.

Definition 10. Global Euclidean distance limit, $\mathcal{R}(a, b)$. Given a pair of DT coefficients (a, b), we define the global Euclidean distance limit $\mathcal{R}(a, b)$ as the minimum between the distance limits induced by both Type 1 and 2 errors. Therefore, $\mathcal{R}(a, b) = \min(\mathcal{R}_1(a, b), \mathcal{R}_2(a, b))$.

$\mathcal{R}(a, b)$ represents the maximal achievable Euclidean distance when growing topologically correct discrete discs. Equivalently, given a pair of DT coefficients (a, b), for any discrete distance value D such that $R_{\max}(D) \leq \mathcal{R}(a, b)$, no topological error (of Type 1 or Type 2) occurs in the discrete disc of radius D. In Fig. 6(B), $(\mathcal{R}(a, b))^2$ is plotted for each DT coefficients pair such that $a \leq 10$.

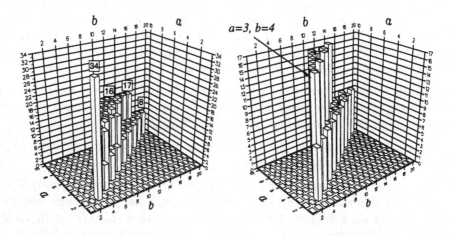

Fig. 6. A:(left) Distance limit induced by Type 2 error. B:(right) Global Euclidean distance limit for the correctness of the EDT.

Note that, for large values of a and b, the limit induced by the Type 2 error dominates. Type 1 error only dominates for the smallest possible values of (a, b) (*i.e.*, $a=2$, $b=3$). For any other pair, Type 2 error dominates. The following table summarises the distance limits obtained for some values of the DT coefficients (a, b).

a	b	D	D'	$R_{max}(D)$	$\mathcal{R}(a, b)$	Type
2	3	6	6	$\sqrt{9}$	$\sqrt{8}$	1
3	4	12	13	$\sqrt{18}$	$\sqrt{17}$	1 and 2
3	5	9	10	$\sqrt{9}$	$\sqrt{8}$	2
5	7	21	22	$\sqrt{18}$	$\sqrt{17}$	2
9	11	33	36	$\sqrt{18}$	$\sqrt{16}$	2

In summary, the results in Sections 3.1 and 3.2 lead us to the characterisation of a Euclidean distance limit $\mathcal{R}(a, b)$ for any pair of DT coefficients (a, b). The Euclidean distance limit can readily give a discrete distance limit via the definitions of R_{max} and R_{min} (see Definition 4).

3.3 Global Euclidean distance limit and optimal DT coefficients

Our aim now is to determine, whether an optimal pair exists among all valid pairs of DT coefficients. We define optimality here as the smallest integer pair of DT coefficients which guarantees the maximum achievable Euclidean distance limit. Using the results plotted in Fig. 6(B), we could say that, for all pair of DT coefficients such that $a \leq 10$, the pair $(3, 4)$ is a local optimum in the sense that it is the smallest pair of DT coefficients that leads to a (local) maximum Euclidean distance limit (*i.e.*, $\mathcal{R}(3, 4) = \sqrt{17}$). In order to extend this result to any pair of DT coefficients, we will use an analytical approach rather than the geometric approach which was used previously.

As suggested in Lemma 5 and Definition 6, an analytical Euclidean distance limit induced by Type 1 errors can be estimated by $R_{max}(ab)$. Since this limit increases with the values of (a, b), we pointed out earlier that Type 2 error dominates for greater values of DT coefficients. Hence, we will mainly concentrate on an analytical study of Type 2 errors and finally combine the result with those of the previous study of Type 1 errors. The result of this study can be stated as follows.

Theorem 11. *Euclidean distance limit and optimal DT coefficients Considering the chamfer distance $d_{a,b}$ as a discrete distance, the maximal error-free Euclidean distance achievable is $\sqrt{17}$ and the smallest integer pair of DT coefficients that achieves this limit is $(a, b) = (3, 4)$.*

We introduce the idea behind the proof of Theorem 11 (full details of the proof of this result can be found in reference [7]). According to the conditions in (2) given by $0 < a < b < 2a$, a pair of DT coefficients (a, b) is valid if and only if the integer coordinates (a, b) in the plane (x, y) lie in the positive quadrant of the plane (*i.e.*, $x \geq 0$ and $y \geq 0$) and between the lines $y = x$ and $y = 2x$. Moreover, given a pair of integer values (a, b) satisfying (2), the only integer pair of DT coefficients which is exactly on the line $y = \frac{b}{a}x$ is the pair (a, b) itself since, by definition, $\gcd(a, b) = 1$. In Fig. 7, the valid pairs (a, b) such that $a \leq 10$ and $b \leq 11$ are represented as dots (o) in the plane (x, y).

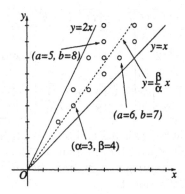

Fig. 7. Representation of the valid pairs of DT coefficients for the proof of Theorem 11.

The approach for deriving an analytical expression of the Euclidean distance limit for Type 2 error (*i.e.*, $\mathcal{R}_2(a, b)$) is made by decomposing the region of the plane (x, y) delimited by the lines $y = x$ and $y = 2x$, by a line $y = \frac{\beta}{\alpha}x$, where (α, β) is an integer pair that matches the conditions for being a pair of DT coefficients (see Fig. 7). For each such pair (α, β), we characterise a Euclidean distance limit in each sub-region of the plane (x, y) delimited by the lines $y = x$, $y = \frac{\beta}{\alpha}x$ and $y = 2x$. Given a valid pair (α, β), and for any pair of DT coefficients (a, b) different from (α, β), two cases are possible, (i): $\frac{b}{a} < \frac{\beta}{\alpha}$ or (ii): $\frac{b}{a} > \frac{\beta}{\alpha}$. Case (i) includes the valid integer points in the sub-region below $y = \frac{\beta}{\alpha}x$ and above $y = x$, whereas Case (ii) includes the valid integer points in the sub-region above $y = \frac{\beta}{\alpha}x$ and below $y = 2x$. The Euclidean distance limit $\mathcal{R}_2(a, b)$ is to be investigated for the two sub-regions separately and we will refer to this as $\mathcal{R}_{\inf}(\alpha, \beta)$ and $\mathcal{R}_{\sup}(\alpha, \beta)$ for cases (i) and (ii) respectively. In the example illustrated in Fig. 7, $\alpha = 3$, $\beta = 4$. Then, for instance, $\mathcal{R}_2(6, 7)$ will include the Euclidean distance limit $\mathcal{R}_{\inf}(3, 4)$, since $\frac{7}{6} < \frac{4}{3}$. Similarly, $\mathcal{R}_2(5, 8)$ will include the Euclidean distance limit $\mathcal{R}_{\sup}(3, 4)$, since $\frac{8}{5} > \frac{4}{3}$.

Hence, given a pair of DT coefficients (a, b), the Euclidean distance limit for Type 2 errors induced by (a, b) (*i.e.*, $\mathcal{R}_2(a, b)$) will result from a combination of all Euclidean distance limits induced by the pairs (α, β) in the following way.

$$\mathcal{R}_2(a, b) = \min \left(\min_{\{(\alpha, \beta) / \frac{b}{a} < \frac{\beta}{\alpha}\}} (\mathcal{R}_{\inf}(\alpha, \beta)), \min_{\{(\alpha, \beta) / \frac{b}{a} > \frac{\beta}{\alpha}\}} (\mathcal{R}_{\sup}(\alpha, \beta)) \right)$$

Proof: Given a pair of DT coefficients (a, b) and an integer pair (α, β) such that $0 < \alpha < \beta < 2\alpha$ and $\gcd(\alpha, \beta) = 1$, we consider two cases: (i) $\frac{b}{a} < \frac{\beta}{\alpha}$ and, (ii)

$\frac{b}{a} > \frac{\beta}{\alpha}$. It will become apparent that the equality case represents a Type 1 error and, therefore, as noted earlier, will not be studied here.

An investigation of cases *(i)* and *(ii)* leads to analytical expressions for the Euclidean distance limits induced by each case (*i.e.*, $\mathcal{R}_{\inf}(\alpha, \beta)$ and $\mathcal{R}_{\sup}(\alpha, \beta)$, respectively). Details for the development of such expressions can be found in reference [7]. Euclidean distance limits induced by all possible values of (α, β) are summarised in the table below (in this case, $\lceil x \rceil$ is the smallest integer strictly greater than x).

	$\mathcal{R}_{\inf}(\alpha, \beta)$	$\mathcal{R}_{\sup}(\alpha, \beta)$
$\frac{\beta}{\alpha} < \sqrt{2}$	β	$\sqrt{\left(\alpha + \left\lceil \frac{2\alpha^2 - \beta^2}{2(\beta - \alpha)} \right\rceil\right)^2 + \alpha^2}$
$\frac{\beta}{\alpha} > \sqrt{2}$	$\sqrt{\left(\beta + \left\lceil \frac{\beta^2 - 2.\alpha^2}{2(2.\alpha - \beta)} \right\rceil\right)^2 + \left\lceil \frac{\beta^2 - 2.\alpha^2}{2(2.\alpha - \beta)} \right\rceil^2}$	$\alpha\sqrt{2}$

We can now list the values of the limits for the smallest possible values of α and β. The following table summarises the Euclidean distance limit values obtained when comparing $\frac{b}{a}$ with $\frac{\beta}{\alpha}$ with the first possible values of α and β.

α	β	$\mathcal{R}_{\inf}(\alpha, \beta)$	$\mathcal{R}_{\sup}(\alpha, \beta)$
2	3	$\sqrt{17}$	$\sqrt{8}$
3	4	$\sqrt{16}$	$\sqrt{34}$
3	5	$\sqrt{97}$	$\sqrt{18}$
4	5	$\sqrt{25}$	$\sqrt{80}$
4	7	$\sqrt{337}$	$\sqrt{32}$

Now, each possible combination of a, b, α and β creates an increasing sequence when ordered such that $\alpha^2 + \beta^2$ increases (*e.g.*, the expression of $\mathcal{R}_{\inf}(\alpha, \beta)$ when $\frac{\beta}{\alpha} > \sqrt{2}$ leads to the increasing sequence $\sqrt{17}$, $\sqrt{97}$, $\sqrt{337}$, $\cdots$). Therefore, all possible Euclidean distance limits will be obtained as soon as we obtain a limit value for any range of $\frac{b}{a}$. Using the two first lines in the previous table, we deduce that, for $\frac{3}{2} < \frac{b}{a} < 2$, $\mathcal{R}_2(a, b) = \sqrt{8}$; for $\frac{b}{a} = \frac{3}{2}$, $\mathcal{R}_2(a, b) = \sqrt{34}$; for $\frac{4}{3} \leq \frac{b}{a} < \frac{3}{2}$, $\mathcal{R}_2(a, b) = \sqrt{17}$; for $1 < \frac{b}{a} < \frac{4}{3}$, $\mathcal{R}_2(a, b) = \sqrt{16}$, which fits exactly the results shown in Fig. 6(A).

As pointed out earlier, $\mathcal{R}_1(a, b)$ increases with the values of the DT coefficients. Hence, clearly $\mathcal{R}(a, b) = \mathcal{R}_2(a, b)$ for any pair of DT coefficients $(a, b) \neq (2, 3)$. Now, $\mathcal{R}_2(2, 3) = \mathcal{R}_{\sup}(3, 4) = \sqrt{34}$, since $\frac{3}{2} > \frac{4}{3}$. From the result of characterisation of Type 1 error, we obtain $\mathcal{R}_1(2, 3) = \sqrt{8}$. Therefore, $\mathcal{R}(2, 3) = \min(\mathcal{R}_1(2, 3), \mathcal{R}_2(2, 3)) = \sqrt{8}$. Hence, the maximal Euclidean distance achievable is $\max_{\{(a, b)\}} \mathcal{R}(a, b) = \sqrt{17}$. Clearly, the first pair (a, b) which realises this maximum is $(a, b) = (3, 4)$. Hence, Theorem 11 holds. $\square$

In summary, we have extended the results derived from the geometrical study presented in Section 3. Theorem 11 states that, for any DT coefficients (a, b) such that $\frac{4}{3} \leq \frac{b}{a} < \frac{3}{2}$, the topological order is preserved in any discrete disc of radius D such that $R_{\max}(D) < \sqrt{17}$. In the design of algorithms which require chamfer distances, it is wise to maintain small values of the discrete distances computed. In this context, the minimum DT coefficients for achieving the global upper bound of $\sqrt{17}$ is $(a, b) = (3, 4)$.

4 Conclusion

The problem of approximating continuous distances by discrete ones was considered. We formally characterised topological errors which occur during the mapping of distances from the discrete to the continuous space. Distance limits up to which these errors are guaranteed not to occur were derived for any pair of DT coefficients. Among all DT coefficients, an optimal integer pair was characterised and shown analytically to correspond to a global optimum.

As by-product of this study, we obtained results which give, without the need of enumeration, all possible decompositions of a discrete distance value into a combination of moves on a shortest path on the grid. Among applications, such results can readily be used for the development of exact Euclidean distance mapping algorithms (see *e.g.*, [8]).

References

1. G. Borgefors. Distance transformations in arbitrary dimensions. *Comput. Vision, Graphics and Image Process.*, 27:321–345, 1984.
2. G. Borgefors. Distance transformations in digital images. *Comput. Vision, Graphics and Image Process.*, 34:344–371, 1986.
3. S. Forchhammer. *Representation and data compression of two-dimensional graphical data*. PhD thesis, Technical University of Denmark, Denmark, 1988.
4. S. Forchhammer. Euclidean distances from chamfer distances for limited distances. In *6th Scandinavian Conf. on Image Analysis*, pp. 393–400, Oulu, Finland, 1989.
5. C. J. Hilditch. Linear skeletons from square cupboards. In B. Meltzer and D. Mitchie, Eds, *Machine Intelligence*, 4:403–420. Edinburgh Univ. Press, 1969.
6. C. J. Hilditch and D. Rutovitz. Chromosome recognition. *Annals of the New York Academy of Sciences*, 157:339–364, 1969.
7. S. Marchand-Maillet and Y. M. Sharaiha. Topologically-correct continuous to discrete distance mapping. T.R. SWP9601/OR, Imperial College, London, 1996.
8. S. Marchand-Maillet and Y. M. Sharaiha. A graph-theoretic algorithm for the exact solution of the Euclidean distance mapping. In *10th Scandinavian Conf. on Image Analysis*, pp. 221–228, Lappeenranta, Finland, 1997.
9. U. Montanari. Continuous skeletons from digitized images. *Journal of the ACM*, 16:534–549, 1969.
10. A. Rosenfeld and J. L. Pfaltz. Distances functions on digital pictures. *Pattern Recognition*, 1:33–61, 1968.
11. J. Sylvester. Mathematical questions with their solutions. *Educational Times*, 41, 1884.
12. E. Thiel and A. Montanvert. Chamfer masks: discrete distance functions, geometrical properties and optimization. In *11th Int. Conf. on Pattern Recognition*, pp. 244–247, The Hague, The Netherlands, 1992.

From Principles to Applications

Homotopy in 2-dimensional Digital Images

Rémy MALGOUYRES

GREYC, ISMRA
6, bd du Maréchal Juin 14050 CAEN cedex FRANCE,
e-mail : Malgouyres@greyc.ismra.fr

Abstract. We recall the basic definitions concerning homotopy in 2D Digital Topology, and we set and prove several results concerning homotopy of subsets. Then we introduce an explicit isomorphism between the fundamental group and a free group. As a consequence, we provide an algorithm for deciding whether two closed path are homotopic.

Key words: Digital topology, homotopy, fundamental group.

Introduction

Homotopy in the framework of Digital Topology is an important question in the field of Image Analysis. In particular, T. Y. Kong introduced a notion of the digital fundamental group in the 3−dimensionnal digital Euclidian space ([1]), and in a more genaral framework ([2]). The purpose of this paper is to study the corresponding notion in the 2−dimensional digital space. First we recall the basic definitions concerning homotopy in 2D Digital Topology. Then we set and prove several results which are required in the sequel concerning homotopy of subsets. Finally we introduce an explicit isomorphism between the fundamental group of any object with m holes and and the free group with m generators. As a consequence, we provide an algorithm for deciding whether two closed path are homotopic or not in a given arbitrary object. The computational complexity of this algorithm is the sum of the lengths of the two considered paths.

1 Basic definitions and notations

If X is a subset of $\mathbb{Z}^2$, we denote $\overline{X} = \mathbb{Z}^2 \backslash X$ the complement of X. In this paper, we shall consider only finite subsets X of $\mathbb{Z}^3$. For $x = (i,j) \in \mathbb{Z}^2$, we consider the two following neighborhoods:

$N_4(x) = \{y = (i',j') \in \mathbb{Z}^2 \ / \ |i - i'| + |j - j'| = 1\}$;

$N_8(x) = \{y = (i',j') \in \mathbb{Z}^2 \ / \ max(|i - i'|, |j - j'|) = 1\}$.

Let $n \in \{4, 8\}$. Two points x and y of $\mathbb{Z}^2$ are said to be n−adjacent if $y \in N_n(x)$. This n−adjacency relation defines a graph structure on $\mathbb{Z}^2$, called the n−adjacency graph. For any subset X of $\mathbb{Z}^2$, $n-connected\ components$ of X are connected components of the subgraph of the n−adjacency graph induced by X. The set X is said to be $n-connected$ if it has a single n−connected component. As usual, when we analyze a set $X \subset \mathbb{Z}^3$ using an n−connectivity type with

$n \in \{4, 8\}$, we analyze $\overline{X}$ with a different $\overline{n}$–connectivity with $\overline{n} = 12 - n$. In the sequel, we consider $(n, \overline{n}) \in \{(4, 8), (8, 4)\}$. An n–hole in $X \subset \mathbb{Z}^2$ is a bounded $\overline{n}$–connected component of $\overline{X}$. A *finite n–path* is a finite sequence $(x_0, \ldots, x_p)$ such that for $i \in \{1, \ldots, p\}$ the point x_{i-1} is n–adjacent or equal to x_i. Such a finite n–path is said to be *closed* if $x_0 = x_p$. An *infinite n–path* is a sequence $(x_i)_{i \in \mathbb{N}}$ such that for $i \in \mathbb{N}^*$ the point x_{i-1} is n–adjacent or equal to x_i. Such an infinite n–path is called *simple* if $i \neq j \implies x_i \neq x_j$. If π is a finite or infinite n–path of $\mathbb{Z}^2$, we denote by π^* the set of the points of π. We also denote by $\pi * \pi'$ the concatenation of two finite n–paths π and π'. Given an n–path $\pi = (x_0, \ldots, x_p)$, we denote by π^{-1} the n–path $(x_p, \ldots, x_0)$.

Now we need to introduce the *n–homotopy relation* between n–paths. Let us consider $X \subset \mathbb{Z}^2$ and two points $B \in X$ and $B' \in X$. We also consider $A^n_{B,B'}(X)$ the set of all closed n–paths $\pi = (x_0, \ldots, x_p)$ which are included in X and such that $x_0 = B$ and $x_p = B'$. First we introduce the notion of an *elementary deformation*. Two finite n–paths $\pi \in A^n_{B,B'}(X)$ and $\pi' \in A^n_{B,B'}(X)$ are said to be *the same up to an elementary deformation (with fixed extremities)* if they are of the form $\pi = \pi_1 * \gamma * \pi_2$ and $\pi' = \pi_1 * \gamma' * \pi_2$, the n–paths γ and γ' having the same extremities and being both included in a common unit square. Now, the two n–paths $\pi \in A^n_{B,B'}(X)$ and $\pi' \in A^n_{B,B'}(X)$ are said to be *n–homotopic (with fixed extremities)* in X if there exists a finite sequence of n–paths $\pi = \pi_0, \ldots, \pi_m = \pi'$ of $A^n_{B,B'}(X)$ such that for $i = 1, \ldots, m$ the n–paths π_{i-1} and π_i are the same up to an elementary deformation (with fixed extremities).

We denote $A^n_B = A^n_{B,B}$ The homotopy relation defines an equivalence relation on $A^n_B(X)$, and we denote by $\Pi^n_1(X)$ the set of equivalence classes of this equivalence relation. The concatenation of closed n–paths is compatible with the homotopy relation, hence it defines an operation on $\Pi^n_1(X)$, and this operation provides $\Pi^n_1(X)$ with a group structure. We call this group the *n–fundamental group of X*. The n–fundamental group defined using a point B' as the based point is isomorphic to the n–fundamental group defined using a point B as the based point.

Now we consider n–connected sets $X \subset Y \subset \mathbb{Z}^2$. First we observe that a closed n–path in X is also a closed n–path in Y. Moreover, two n–homotopic closed n–paths in X are also n–homotopic in Y. These two properties enables us to define a canonical morphism $i_* : \Pi^n_1(X) \longrightarrow \Pi^n_1(Y)$ which is called the morphism induced by the inclusion $i : X \longrightarrow Y$.

Now we must introduce an algebraic notion called *the free group with m generators*. Let $\{a_1, \ldots, a_m\} \cup \{a_1^{-1}, \ldots, a_m^{-1}\}$ be an alphabet with $2m$ distinct letters, and let L_m be the set of the all words over this alphabet (i.e. finite sequences of letters of the alphabet). We say that two words $w \in L_m$ and $w' \in L_m$ are *the same up to an elementary simplification* if, either w can be obtained from w' by inserting in w' a sequence of the form $a_i a_i^{-1}$ or a sequence of the form $a_i^{-1} a_i$ with $i \in \{1, \ldots, m\}$, or w' can be obtained from w by inserting in w a sequence of the form $a_i a_i^{-1}$ or a sequence of the form $a_i^{-1} a_i$ with $i \in \{1, \ldots, m\}$. Now, two words $w \in L_m$ and $w' \in L_m$ are said to be *free equivalent* if there is a finite sequence $w = w_1, \ldots, w_k = w'$ of words of L_m such that for $i = 2, \ldots, k$ the word w_{i-1} and w_i are the same up to an elementary simplification. This defines an equivalence relation on L_m, and we denote by $\mathcal{F}_n$ the set of equivalence classes of this equivalence relation. If $w \in L_m$, we denote by $\overline{w}$ the class of w under the free equivalence relation. The concatenation of words defines an operation on $\mathcal{F}_n$ which provides $\mathcal{F}_n$ with a group structure. The group thus defined is called the *free group with m generators*. We denote by 1_m the unit element of $\mathcal{F}_m$, which is equal to $\overline{w}$ where w is the empty word. The only result which we shall admit on the free group is the classical result that if a word $w \in L_n$ is such that $\overline{w} = 1_m$ and w is not the empty word, then there exists in w two successive letters $a_i a_i^{-1}$ or $a_i^{-1} a_i$ with $i \in \{1, \ldots, m\}$. This remark leads to an immediate algorithm to decide whether a word $w \in L_n$ is such that $\overline{w} = 1_m$.

2 On the fundamental group of subsets

In this section, we state and prove some results relative to inclusion of sets and the fundamental group. First we set a definition :

Definition 1. Let $X \subset \mathbb{Z}^2$ and $x \in X$. The point x is called $n-simple$ if the number of $n-$connected components of $N_8(x) \cap X$ which are $n-$adjacent to x is equal to 1, and $N_{\overline{n}}(x) \cap \overline{X} \neq \emptyset$.

Observe that if $x \in X$ is such that $N_n(x) \cap X$ is nonempty such that $N_{\overline{n}}(x) \cap \overline{X} \neq \emptyset$, then x is $n-$simple if and only if the number of $\overline{n}-$connected components of $N_8(x) \cap \overline{X}$ which are $\overline{n}-$adjacent to x is equal to 1.

Let $X \subset Y \subset \mathbb{Z}^2$. The set X is said to be *lower $n-$homotopic to Y* if X can be obtained from Y be deleting sequentially $n-$simple points. In this case the set Y is called *upper $n-$homotopic to X*. Finally, the set X and Y are called $n-$homotopic if there exists a finite sequence $X_0, \ldots, X_m \subset \mathbb{Z}^3$ of sets such that $X = X_0$ and $Y = X_m$ and for $i = 1, \ldots, m$ the set X_{i-1} is either lower $n-$homotopic or upper $n-$homotopic to X_i.

Lemma 2. *Let $X \subset \mathbb{Z}^2$, let $B, B' \in X$ and $x \in X$ an $n-simple$ point which is distinct from B and B'. Then if two $n-paths$ π and π' of $A_{B,B'}^n (X \backslash \{x\})$ are $n-homotopic$ (with fixed extremities) in X, they are $n-homotopic$ in $X \backslash \{x\}$.*

Proof: First, if $c = (x_0, \ldots, x_p)$ is an $n-$path in X such that $x_0 \neq x$ and $x_p \neq x$, we define an $n-$path $P(c)$ as follows: For any maximal sequence $\sigma = (x_k, \ldots, x_l)$ with $0 \leq k \leq l \leq p$ of points of c such that for $i = k, \ldots, l$ we have $x_i \neq x$, we define $c(\sigma) = \sigma$. For any maximal sequence $\sigma = (x_k, \ldots, x_l)$ with $1 \leq k \leq l < p$ of points of c such that for $i = k, \ldots, l$ we have $x_i = x$, we define $c(\sigma)$ as equal to the shortest $n-$path in $N_8(x) \cap X$ from x_{l-1} to x_{k+1}. Now, $P(c)$ is the concatenation of all $c(\sigma)$ for all maximal sequence $\sigma = (x_k, \ldots, x_l)$ of points of c such that either for $i = k, \ldots, l$ we have $x_i \neq x$ or for $i = k, \ldots, l$ we have $x_i = x$.

Now, it is sufficient to prove that if π and π' are two elements of $A^n_{B,B'}(X)$ and are the same up to an elementary deformation, the two n-paths $P(\pi)$ and $P(\pi')$ also are the same up to an elementary deformation. Hence we assume π and π' are of the form $\pi = \pi_1 * \gamma * \pi_2$ and $\pi' = \pi_1 * \gamma' * \pi_2$, the n-paths γ and γ' having the same extremities and being both included in a common unit square S. Without loss of generality, we assume that $x \in S$. Let $\pi_1 = (x_{1,0}, \ldots, x_{1,k_1})$ and $\pi_2 = (x_{2,0}, \ldots, x_{2,k_2})$. We denote by α_1 the shortest n-path in $N_8(x) \cap X$ from the last point of π_1 to S, and we denote by α_2 the shortest n-path in $N_8(x) \cap X$ from S to the first point of π_2. We denote $\alpha_1 = (y_{1,0}, \ldots, y_{1,k_1})$ and $\alpha_2 = (y_{2,0}, \ldots, y_{2,k_2})$. Finally, we define $\delta = (y_{1,k_1}) * \gamma * (y_{2,0})$, and $\delta' = (y_{1,k_1}) * \gamma' * (y_{2,0})$. Now we have $P(\pi) = (P(\pi_1) * \alpha_1) * P(\delta) * (\alpha_2 * P(\pi_2))$ and $P(\pi') = (P(\pi_1) * \alpha_1) * P(\delta') * (\alpha_2 * P(\pi_2))$. Since $P(\delta)$ and $P(\delta')$ have the same extremities and are both included in the unit square S, the n-paths $P(\pi)$ and $P(\pi')$ are the same up to an elementary deformation. $\square$

Corollary 3. *Let $X \subset Y \subset \mathbb{Z}^2$ be such that X is lower n-homotopic to Y. Let $B, B' \in X$. Then if two closed n-paths π and π' of $A^n_{B,B'}(X)$ are n-homotopic (with fixed extremities) in Y, they are n-homotopic in X.*

Lemma 4. *Let $X \subset \mathbb{Z}^2$, let $B, B' \in X$ and $x \in X$ an n-simple point distinct from B and B'. Then any n-path c of $A^n_{B,B'}(X)$ is n-homotopic (with fixed extremities) to an n-path contained in $X \backslash \{x\}$.*

Proof: Let $P(c)$ be the n-path as defined in the proof of Lemma 2. It is easy to see that c is n-homotopic (with fixed extremities) to $P(C)$. $\square$

Corollary 5. *Let $X \subset Y \subset \mathbb{Z}^2$ be such that X is lower n-homotopic to Y, and let $B, B' \in X$. Then any n-path c of $A^n_{B,B'}(Y)$ is n-homotopic (with fixed extremities) to an n-path contained in X.*

Corollary 6. *Let $X \subset Y \subset \mathbb{Z}^2$ be such that X is lower n-homotopic to Y. The morphism $i_* : \Pi^n_1(X) \longrightarrow \Pi^n_1(Y)$ induced by the inclusion map is a group isomorphism.*

The following result is folklore:

Theorem 7. *Let $X \subset Y \subset \mathbb{Z}^2$ be two n-connected sets. Suppose that any $\bar{n}$-connected component of $\overline{X}$ contains exactly one $\bar{n}$-connected component of $\overline{Y}$. Then X is lower n-homotopic to Y.*

3 The noncommutative winding number

In this section, for any $X \subset \mathbb{Z}^2$ with m n-holes and $B \in X$, for any $c \in A^n_B(X)$, we define a word $W \in L_m$. The corresponding element $\overline{W}$ of $\mathcal{F}_m$ is called the *noncommutative winding number of c*. The idea is the following: first we chose a point P_i in the ith n-hole of X. Then we consider a particular infinite simple 4−path π which contains all points $P_1, \ldots, P_m$. By remaining $P_1, \ldots, P_m$ if necessary, we may assume that the order in which the P_i's appear in π is the order of increasing i's. Then we construct the word w following c, adding a symbol a_i or a_i^{-1} to the word we construct each time c crosses the section of π between P_i and P_{i+1}, depending on how c crosses π. For instance, in Figure 1, the noncommutative winding number is equal to $a_2^{-1} a_1 a_2 a_1^{-1}$.

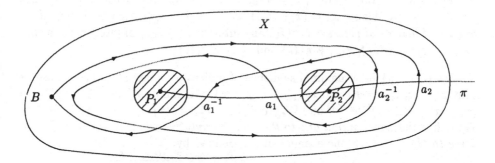

Fig. 1. Example with non commutative winding number equal to $a_2^{-1} a_1 a_2 a_1^{-1}$.

Let us now consider all this more precisely. In the following, X is a subset of $\mathbb{Z}^2$ with m n−holes. We chose a point $P_i = (\alpha_i, \beta_i)$ in the ith n−hole of X. Let $R = [a, b] \times [a', b']$ be a rectangle such that X is contained in R. We denote $X_1 = R \backslash \{P_1, \ldots, P_m\}$.

We construct a particlular infinite simple 4−path $\pi = (y_i)_{i \in \mathbb{N}}$ (see Figure 2) as follows: Let $k_1 = b + 1 - \alpha_1$, and for $j = 2, \ldots, \beta_m - \beta_1 + 1$, let $k_j = k_{j-1} + b - a + 3$. For convenience, we set $k_{\beta_m - \beta_1 + 2} = +\infty$. For $i \in \{0, \ldots k_1\}$ we set $y_i = (\alpha_1 + i, \beta_1)$, and for $k_j < i \leq k_{j+1}$ with $j \geq 2$, we set $y_i = (a + i - k_j, \beta_1 + j - 1)$ if j is even and $y_i = (b - (i - k_j), \beta_1 + j - 1)$ otherwise.

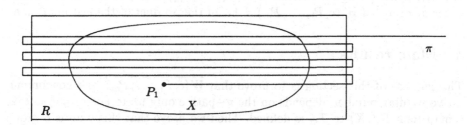

Fig. 2. The 4−path π.

In other words, π is the concatenation of strait line segments, $[(a - 1, \beta_1 + j - 1), (b + 1, \beta_1 + j - 1)]$ or $[(b + 1, \beta_1 + j - 1), (a - 1, \beta_1 + j - 1)]$, depending on the parity of j (except for $j = 1$ for which we have a segment $[(\alpha_1, \beta_1), (b + 1, \beta_1)]$).

By remaining the P_i's if necessary, we may assume that the order in which they appear in the 4−path π is the order of increasing i's.

Now, for $k = 1, \ldots, m$, we denote by e_k the unique integer such that $y_{e_k} = P_k$. For convenience, we denote $e_{m+1} = +\infty$. We denote by I_k the interval of integers $\{e_k, \ldots, e_{k+1}\}$. We also denote $\pi(I_k) = \{y_{e_k}, \ldots, y_{e_{k+1}}\}$.

In the sequel, B is the base point of X and we assume without loss of generality that it has its second coordinate less than β_1. In particular, $B \notin \pi^*$. In the sequel of this section, $c = (x_0, \ldots, x_p)$ is an element of $A_B^n(X_1)$. For $i = 0, \ldots, p$, we denote $x_i = (x_{i,1}, x_{i,2})$. For $k = 1, \ldots, m$, we call *maximal sequence of indices of points* $c^* \cap \pi(I_k)$ any interval $\{i, \ldots, j\}$ of integers such that $\{x_i, \ldots, x_j\} \subset \pi(I_k)$, $x_{i-1} \notin \pi(I_k)$ and $x_{j+1} \notin \pi(I_k)$.

Let $\{i, \ldots, j\}$ be a maximal sequence of indices of points $c^* \cap \pi(I_k)$. Observe that $x_{i,2} - x_{i-1,2} = \pm 1$ and $x_{j+1,2} - x_{j,2} = \pm 1$. We denote $W_{b,i}(c, P_1, \ldots, P_m) = (x_{i,2} - x_{i-1,2}) . (-1)^{(x_{i,2} - \beta_1)}$ and $W_{e,j}(c, P_1, \ldots, P_m) = (x_{j+1,2} - x_{j,2}) . (-1)^{(x_{j,2} - \beta_1)}$. We define the *contribution of i to the noncommutative winding number of c relative to* $\{P_1, \ldots, P_m\}$, the element of $\mathcal{F}_m$ defined by:
$$W_i(c, P_1, \ldots, P_m) = a_k^{\frac{W_{b,i}(c, P_1, \ldots, P_m) + W_{e,j}(c, P_1, \ldots, P_m)}{2}}$$
$W_i(c, P_1, \ldots, P_m)$ is an element of $\mathcal{F}_m$ which is equal either to a_k, or to a_k^{-1}, or to 1_m. For convenience, we denote $W_i(c, P_1, \ldots, P_m) = 1_m$ if i is not the smallest element of a maximal sequence of indices of points of some $c^* \cap \pi(I_k)$ with $k \in \{1, \ldots, m\}$, so that $W_i(c, P_1, \ldots, P_m)$ is defined for any $i \in \{0, \ldots, p\}$.

Definition 8. We call *noncommutative winding number of c relative to* $\{P_1, \ldots, P_m\}$ the element of $\mathcal{F}_m$ defined by:
$$\overline{W(c, P_1, \ldots, P_m)} = \prod_{i=0}^{p} W_i(c, P_1, \ldots, P_m)$$

Observe that, since $\mathcal{F}_m$ is not an abelian group, the order in which the product is defined in this last definition must be respected.

We also define a *word* $W(c, P_1, \ldots, P_m)$ of L_m by the sequence of letters obtained by replacing the element $W_i(c, P_1, \ldots, P_m)$ of $\mathcal{F}_m$ by the corresponding letter a_k or a_k^{-1} if $W_i(c, P_1, \ldots, P_m) \neq 1_m$ in the product of the last definition.

4 Main results

The purpose of this section is to prove that $\overline{W(c, P_1, \ldots, P_m)}$, the noncommutative winding number, depends on the n−path c only up to homotopy, so that a map from $\Pi_1^n(X)$ to $\mathcal{F}_m$ is defined. Then we prove that this map is a group isomorphism.

Theorem 9. *If two n−paths $c, c' \in A_B^n(X_1)$ are n−homotopic (with fixed extremities), then $\overline{W(c, P_1, \ldots, P_m)} = \overline{W(c', P_1, \ldots, P_m)}$*

Proof: We only need to prove this result when c and c' are the same up to an elementary deformation. So, let $c = c_1 * \gamma * c_2$ and $c' = c_1 * \gamma' * c_2$, the n−paths

γ and γ' having the same extremities and being both included in a common unit square S. If γ and γ' are both included in $\pi(I_k)$ with $k \in \{1, \ldots, m\}$, the result is obvious. Hence we may assume that S meets two sets $\pi(I_k)$ and $\pi(I_{k'})$ with $k \neq k'$. Moreover, it is easily seen that the unit square S meets at most two sets of the form $\pi(I_j)$ with $j \in \{1, \ldots, m\}$. Hence the two lower points of S are, say, in $\pi(I_k) \cup \{P_1, \ldots, P_m\}$ and the two upper points of S then are in $\pi(I_{k'}) \cup \{P_1, \ldots, P_m\}$.

We denote $c = (x_0, \ldots, x_p)$, $c' = (x'_0, \ldots, x'_p)$, for $i \in \{0, \ldots, p\}$ $x_i = (x_{i,1}, x_{i,2})$ and for $i \in \{0, \ldots, p'\}$ $x'_i = (x'_{i,1}, x'_{i,2})$. We also denote $\gamma = (x_{i_0}, \ldots, x_{i_1})$ and $\gamma' = (x'_{i'_0}, \ldots, x'_{i'_1})$.

Let us consider the case when $x_{i_0} = x'_{i'_0} \in \pi(I_k)$ and $x_{i_1} = x'_{i'_1} \in \pi(I_{k'})$. Let $d = min(\{j \geq i_0 \ / \ x_j \in \pi(I_{k'})\})$ and $d' = min(\{j \geq i_0 \ / \ x'_j \in \pi(I_{k'})\})$. Let $f = max(\{j \leq i_1 \ / \ x_j \in \pi(I_k)\})$ and $f' = max(\{j \leq i'_1 \ / \ x'_j \in \pi(I_k)\})$. If $(i, \ldots, j)$ is a maximal sequence of indices of points of $c^* \cap \pi(I_k)$ which is included in $\{i_0, \ldots, i_1\}$, if $j \neq d - 1$ and $i \neq f + 1$ we have $W_{b,i}(c, P_1, \ldots P_m) = -W_{e,j}(c, P_1, \ldots P_m)$, hence $W_i(c, P_1, \ldots, P_m) = 1_m$. Similarly, for $(i, \ldots, j)$ a maximal sequence of indices of points of $c'^* \cap \pi(I_k)$ which is included in $\{i_0, \ldots, i'_1\}$, if $j \neq d' - 1$ and $i \neq f' + 1$ we have $W_i(c', P_1, \ldots, P_m) = 1_m$. Hence we consider $(i, \ldots, d-1)$ the maximal sequence of indices of points of $c^* \cap \pi(I_k)$ which contains $d - 1$, and we consider $(i, \ldots, d' - 1)$ the maximal sequence of indices of points of $c'^* \cap \pi(I_k)$ which contains $d' - 1$. We have $W_{e,d-1}(c, P_1, \ldots, P_m) = (-1)^{(x_{d-1,2} - \beta_1)} = (-1)^{(x'_{d'-1,2} - \beta_1)} = W_{e,d'-1}(c', P_1, \ldots, P_m)$. On the other hand, we have $W_{b,i}(c, P_1, \ldots, P_m) = W_{b,i}(c', P_1, \ldots, P_m)$, and therefore $W_i(c, P_1, \ldots, P_m) = W_i(c', P_1, \ldots, P_m)$. Similarly, we can obtain $W_f(c, P_1, \ldots, P_m) = W_{f'}(c', P_1, \ldots, P_m)$ so that $W(c, P_1, \ldots, P_m) = W(c', P_1, \ldots, P_m)$.

The case when x_{i_0} and x_{i_1} lie on the same horizontal line is similar, except in the case, say, γ is included in $\pi(I_k)$ and γ' is not. In this case, we prove that, either $W(c, P_1, \ldots, P_m) = W(c', P_1, \ldots, P_m)$, or the word $W(c', P_1, \ldots, P_m)$ is obtained from $W(c, P_1, \ldots, P_m)$ by inserting $a_k a_k^{-1}$ or $a_k^{-1} a_k$ in $W(c, P_1, \ldots, P_m)$. In both cases, $\overline{W(c, P_1, \ldots, P_m)} = \overline{W(c', P_1, \ldots, P_m)}$. $\square$

Remark 10. From Theorem 9 follows that the map $c \longmapsto \overline{W(c, P_1, \ldots, P_m)}$ from $A_B^n(X)$ to $\mathcal{F}_m$ induces a map $\varphi_X : \Pi_1^n(X) \longrightarrow \mathcal{F}_m$. This map φ_X is a group morphism.

In order to prove that the map φ_X is a group isomorphism, we first need some technical lemmas.

Lemma 11. Let $c \in A_B^n(X_1)$ be such that $W(c, P_1, \ldots, P_m)$ is the empty word. Then c is $n-$homotopic in X_1 to an $n-$path which contains only the point B.

Proof: For convenience, we call $\pi(I_{m+1})$ and $\pi(I_{m+2})$ the $n-$connected components of $X_1 \backslash \pi^*$. We also denote $m' = m + 2$. First of all, since π^* has no $n-$hole, is not bounded, and contains all $n-$holes of X_1, the set $X_1 \backslash \pi^*$ has no $n-$hole.

Hence if we assume that $c \cap \pi^* = \emptyset$, it follows from Theorem 7 and Corollary 5 that c is n-homotopic in $X_1 \backslash \pi^*$ to an n-path which contains only the point B. Now, we shall prove that c is n-homotopic in X_1 to an n-path which contains only the point B by induction on the number $\mathcal{I}(c)$ of maximal intervals $\{i, \ldots, j\}$ such that there exists $k \in \{1, \ldots, m'\}$ with $\{x_i, \ldots, x_j\} \subset \pi(I_k)$. Suppose that the result is true for any $c' \in A_B^n(X_1)$ with $\mathcal{I}(c') \leq h$ with $h \geq 0$, and $W(c, P_1, \ldots, P_m)$ is the empty word. Suppose that c is such that $\mathcal{I}(c) = h + 1$. We distinguish two case :

First case: for all maximal interval $\{i, \ldots, j\}$ of points of $c^* \cap \pi(I_k)$ with $k \in \{1, \ldots, m'\}$ we have: $x_{i-1,2} \neq x_{j+1,2}$ or $\exists k' \in \{1, \ldots, m'\}$ such that $\{x_{i-1,2}, x_{j+1,2}\} \subset \pi(I_{k'})$. We consider the first maximal interval $\{i, \ldots, j\}$ of indices of points of $c^* \cap \pi(I_k)$ with $k \in \{1, \ldots, m\}$ such that $x_{j+1,2} = x_{j,2} - 1$. Since $W_i(c, P_1, \ldots, P_m) = 1_m$ and $x_{i,2} = x_{i-1,2} + 1$, $x_{i,2}$ and $x_{j,2}$ have the same parity, hence they are equal. Therefore, from the assumptions of the first case we are dealing with, there exists $k' \in \{1, \ldots, m'\}$ such that $\{x_{i-1,2}, x_{j+1,2}\} \subset \pi(I_{k'})$. It is easily seen that the horizontal strait line segment $S = [x_{i-1}, x_{j+1}]$ is included in $\pi(I_{k'})$ and that $S \cup \pi(I_k)$ has no n-hole. Hence, from Theorem 7 and Corollary 5 follows that the n-path $c_1 = (x_{i-1}, \ldots, x_{j+1})$ is n-homotopic with fixed extremities to an n-path contained in S. Therefore, c is n-homotopic to an n-path $c' \in A_B^n(X_1)$ with $\mathcal{I}(c') = h$. Furthermore, it is easy to see that $W(c', P_1, \ldots, P_m)$ is the empty word, hence from our induction hypothesis c' is n-homotopic in X_1 to a constant n-path. This completes the proof in the first case.

Second case: there exists a maximal interval $\{i, \ldots, j\}$ of indices of points such that $\{x_i, \ldots, x_j\} \subset \pi(I_k)$ with $k \in \{1, \ldots, m'\}$ such that $x_{i-1,2} = x_{j+1,2}$ and $\exists k' \neq k''$ with $x_{i-1,2} \in \pi(I_{k'})$ and $x_{j+1,2} \in \pi(I_{k''})$. Since $x_{i,2} \neq x_{i-1,2} = x_{j+1,2}$, we have either $k < k'$ and $k < k''$ or $k > k'$ and $k > k''$. We assume for instance that $k' < k'' < k$. let $\{i', \ldots, j'\}$ be the last maximal interval of indices of points such that $\{x_{i'}, \ldots, x_{j'}\} \subset \pi(I_{k''})$ with $k''' \in \{1, \ldots, m'\}$ such that for all $j \leq i'' \leq j'$ $x_{i''} \in \pi(I_{k''}) \cup \cdots \cup \pi(I_{k-1})$. We denote by S_1 and S_2 respectively the two horizontal strait line segments $[(a, x_{j,2}), (b, x_{j,2})]$ and $[(a, x_{j+1,2}), (b, x_{j+1,2})]$. For any $i'' \in \{i' - 1, \ldots, j'\}$ we have : $x_{i''} \in S_1 \cup S_2$. Since $x_{i'-1} \in S_1 \cup S_2$ and $W_{i'}(c, P_1, \ldots, P_m) = 1_m$, we must have $x_{j'+1} \in S_1 \cup S_2$. It is easily seen that the only possibility is that $x_{j'+1,2} = x_{j,2}$ and $x_{j'+1} \in \pi(I_k)$. Hence we can proceed as in the first case : The strait line segment $S = [x_j, x_{j'+1}]$ is contained in $\pi(I_k)$ and in S_1. Since $(S_1 \cup S_2) \cap X_1$ has no n-hole, the n-path $c_1 = (x_j, \ldots, x_{j'+1})$ is n-homotopic with fixed extremities to an n-path contained in S, hence, as in the first case, c is n-homotopic to an n-path $c' \in A_B^n(X_1)$ with $\mathcal{I}(c') = h$ and such that $W(c', P_1, \ldots, P_m)$ is the empty word. $\square$

Now we observe that the definition of $W_i(c, P_1, \ldots, P_m)$ makes sense for a non-closed n-path $c = (x_0, \ldots, x_p)$ and for any maximal sequence $(i, \ldots, j)$ of indices of points of some $c^* \cap \pi(I_k)$ such that $j \neq p$ and $i \neq 0$.

Lemma 12. *Let $k \in \{1, \ldots, m\}$ and $\varepsilon \in \{-1, 1\}$. Then there exists an n-path $c = (x_0, \ldots, x_p)$ from B to a point of $\pi(I_k)$ such that (denoting $x_i = (x_{i,1}, x_{i,2})$ for $i \in \{0, \ldots, p\}$):*

1. For any maximal sequence $(i, \ldots, j)$ of indices of points of some $c^* \cap \pi(I_{k'})$ such that $j \neq p$ and $i \neq 0$ we have: $W_i(c, P_1, \ldots, P_m) = 0$;

2. If $(i, \ldots, p)$ is the maximal sequence of indices of points of $c^* \cap \pi(I_k)$ which contains p we have: $(x_{i,2} - x_{i-1,2}).(-1)^{x_{i,2}-\beta_1} = \varepsilon$.

Proof: We distinguish two cases depending on whether $\pi(I_k)$ is included in a strait line segment or not. We treat for instance the first case, assuming that $\pi(I_k)$ is included in a strait line segment S at the height h. Let S_1 and S_2 be respectively the set of the points of R of height $h - 1$ and $h + 1$. We assume for instance that $h - \beta_1$ is even. Let a_1 be the first coordinate of the leftmost $P_{k'}$ with $k' \in \{1, \ldots, m\}$. we construct an n–path c_1' by following a horizontal line from B to a point M_1 having $a_1 - 1$ as its first coordinate, and then a vertical line from the point M_1 to the point $M_2 = (a_1 - 1, h + 1) \in S_2$. Since $(S \cup S_2) \cap X_1$ is n–connected, there is an n–path c_2' contained in $(S \cup S_2) \cap X_1$ from M_2 to a point of $\pi(I_k)$. Let $c_1 = (x_0, \ldots, x_p)$ be the concatenation of c_1' and c_2'. We denote $x_i = (x_{i,1}, x_{i,2})$ for $i \in \{0, \ldots, p\}$. Then it is clear that for any maximal sequence $(i, \ldots, j)$ of indices of points of some $c_1^* \cap \pi(I_{k'})$ such that $j \neq p$ and $i \neq 0$ we have: $W_i(c_1, P_1, \ldots, P_m) = 0$. Moreover, if $(i, \ldots, p)$ is the maximal sequence of indices of points of $c_1^* \cap \pi(I_k)$ which contains p we have: $x_{i-1} \in S_2$ so that $(x_{i,2} - x_{i-1,2}).(-1)^{x_{i,2}-\beta_1} = (-1).(h - \beta_1) = -1$.

Now, by considering b_1 the first coordinate of the rightmost point P_i for $i = 1, \ldots, m$, we construct an n–path c_2 satisfying the conditions *1.* and *2.* with $\varepsilon = 1$. $\square$

Lemma 13. *Let $c = (x_0, \ldots, x_p)$ be a closed n–path with $x_0 = x_p \in \pi(I_k)$ with $k \in \{1, \ldots, m\}$, such that or any maximal sequence $(i, \ldots, j)$ of indices of points of some $c^* \cap \pi(I_{k'})$ such that $j \neq p$ and $i \neq 0$ we have: $W_i(c, P_1, \ldots, P_m) = 0$. Moreover we assume that (denoting $x_i = (x_{i,1}, x_{i,2})$ for $i \in \{0, \ldots, p\}$) if $j' = -1 + \min\{0 \leq i \leq p \ / \ x_i \notin \pi(I_k)\}$ and $i' = 1 + \max\{0 \leq i \leq p \ / \ x_i \notin \pi(I_k)\}$, $(x_{i',2} - x_{i'-1,2}).(-1)^{x_{i',2}-\beta_1} = -(x_{j'+1,2} - x_{j',2}).(-1)^{x_{j',2}-\beta_1}$.*
Then c is n–homotopic (with fixed extremities) in X_1 to a constant n–path.

Proof: We denote $\varepsilon = (x_{i',2} - x_{i'-1,2}).(-1)^{x_{i',2}-\beta_1} = -(x_{j'+1,2} - x_{j',2}).(-1)^{x_{j',2}-\beta_1}$. Let $c' = (x_0', \ldots, x_{p'}')$ be the n–path from B to a point $x_{p'}'$ of $\pi(I_k)$ given by Lemma 12. Since $\pi(I_k)$ is n–connected, we may assume that $x_{p'}' = x_0$. The n–path c is n–homotopic in X_1 to $c'^{-1} * c' * c * c'^{-1} * c'$. Now, from Lemma 11, the n–path $c' * c * c'^{-1}$ is n–homotopic in X_1 to a constant n–path. Hence c is n–homotopic in X_1 to a constant n–path. $\square$

Theorem 14. *The map $\varphi_X : \Pi_1^n(X) \longrightarrow \mathcal{F}_m$ defined in Remark 10 is a group isomorphism.*

Proof: First we prove that φ_X is one to one, i.e. that $\overline{W(c, P_1, \ldots, P_m)} = 1_m$ implies that c is homotopic in X to a constant n–path. We already have proved it (Lemma 11) in the case $W(c, P_1, \ldots, P_m)$ is the empty word. Now, we prove our result by induction on the length of the word $W(c, P_1, \ldots, P_m)$. Let us assume that $c = (x_0, \ldots, x_p) \in A_B^n(X)$ is such that $W(c, P_1, \ldots, P_m)$ contains a sequence $a_k a_k^{-1}$ or $a_k^{-1} a_k$, say $a_k a_k^{-1}$, and that the result is true for shorter n-paths. Let $(i, \ldots, j)$ and $(i' \ldots, j')$ be the corresponding respective maximal sequences of indices of points of $c^* \cap \pi(I_k)$ such that $W_i(c, P_1, \ldots, P_m) = a_k$ and $W_{i'}(c, P_1, \ldots, P_m) = a_k^{-1}$. Necessarily, $W_{e,j}(c, P_1, \ldots, P_m) = 1$ and $W_{b,i'}(c, P_1, \ldots, P_m) = -1$. Let c' be an n-path from $x_{i'}$ to x_j which is contained in $\pi(I_k)$. We denote $c_1 = (x_0, \ldots, x_j)$, $c_2 = (x_j, \ldots, x_{i'})$ and $c_3 = (x_{i'}, \ldots, x_p)$ Then $c = c_1 * c_2 * c_3$ is n-homotopic in X_1 to $c_1 * c_2 * c' * c'^{-1} * c_3$. Now, from Lemma 13, $c_2 * c'$ is n-homotopic with fixed extremities to a constant n-path, so that c is n-homotopic to $c_1 * c'^{-1} * c_3$. Since $W(c_1 * c'^{-1} * c_3, P_1, \ldots, P_m)$ is the word obtained from $W(c, P_1, \ldots, P_m)$ by removing a sequence $a_k a_k - 1$, we apply our induction hypothesis to $c_1 * c'^{-1} * c_3$, so that it is n-homotopic in X_1 to a constant n-path. Therefore c also is. From Theorem 7 and Corollary 5, c is then n-homotopic in X to a constant n-path.

Now, for proving that φ_X is onto, we only observe that by applying twice Lemma 12 we obtain, for any $k \in \{1, \ldots, m\}$, an n-path c such that $W(c, P_1, \ldots, P_m) = a_k$. □

Corollary 15. *There is an algorithm for deciding whether two n-paths c and c' of $A_B^n(X)$ are n-homotopic in X whose complexity is the sum of the lengths of c and c'.*

Conclusion

Now, besides the characterization of low homotopy of sets using the fundamental group, we have a complete presentation of the fundamental group of any object in a 2-dimensionnal digital image. Moreover, since the word problem (i.e. the problem of knowing whether a word in the generators is trivial or not) has a simple solution in a free group, we can decide whether two closed path are homotopic or not.

Of course, the same problem exists in three dimensions, but it seems rather more complicated. First of all, we do not have a good characterization of low homotopy of sets, and this is probably the first problem to solve. Then we know that the word problem is not decidable in general ([3]). Hence a first step is may be to treat the fundamental group of surfaces since we know that, in the continuous framework, the word problem is decidable for the surface groups.

References

[1]   T. Y. Kong, A Digital Fundamental Group *Computer and Graphics*, **13**, pp 159-166, 1989.

[2]   T. Y. Kong, Polyhedral Analogs of Locally Finite Topological Spaces *R. M. Shortt editor, General Topology and Applications: Proceedings of the 1988 Northeast Conference, Middletown, CT (USA), Lecture Notes in Pure and Applied Mathematics*, **123**, pp 153-164, 1990.

[3]   A. Tarski, Indecidability of group theory, *Journal of Symbolic Logic*, **14**, pp 76-77, 1949.

Set Manipulations of Fractal Objects
Using Matrices of IFS

Joëlle Thollot

LIGIM - Bât 710 - Université Claude Bernard

43, bd du 11 Novembre 1918

69622 VILLEURBANNE Cedex

Tel: 04.72.44.80.00 ext. 42.74 Fax: 04.72.43.13.12

E-mail: jthollot@ligim.univ-lyon1.fr

Laboratoire d'Informatique Graphique Image et Modélisation

ABSTRACT. One of the major problems in geometric modeling is the control of shape construction. Indeed, one should be able to construct geometrical forms by combining or manipulating simple entities. This problem is even more important when we deal with fractal geometry. In this paper, we propose some methods for increasing the modeling capabilities of fractal shape constructions. We propose an extension of the IFS model based on the definition of matrices of IFS that provides a constructive approach of fractal shapes.

KEY WORDS. *Fractals, geometric modeling, IFS, matrices of IFS.*

1   Introduction

The aim of geometric modeling is to model forms by manipulating or combining well-known shapes such as circles and boxes. In order to be convenient such a model should allow easy control and manipulation of the final shape. When dealing with fractal geometric modeling, the control on fractal figures is not so easily achieved as with classical smooth ones. The difficulty comes from the fact that fractal shapes are usually generated with iterative or recursive procedures.

In order to have a better control on the shape, we wish to develop a *fractal modeler*. Such a modeler should be constituted of a set of basic shapes, a set of unary operations (shape modifications) and a set of binary operations (shape combinations).

Our work focuses on set operations based on the IFS model. This approach has been inspired by constructive solid geometry (CSG). This technique is classical in geometric modeling. It permits to build complex objects using set operations (union, intersection, ...). In fractal modeling such operations are possible using IFS matrices which are a way to generate a wide class of fractal shapes, not necessarily self-similar. These matrices can be combined using certain operations to yield complex fractal shapes in a constructive way.

2 IFS modeling

Fractals are characterized by their property of self-similarity. BARNSLEY's idea [Bar88] consists in using this property in order to encode the different parts of the fractal by contractive operators. An iterative scheme based on these operators is applied. This scheme transforms any initial shape into a union of its reduced copies. The contraction property of the operators is used to prove the existence of a unique invariant set named the attractor. Below, we summarize the basic notions of IFS theory and introduce the notation used in the remainder of the paper.

2.1 Definitions

An IFS-based modeling system is defined by a triple $(\mathcal{X}, d, \mathcal{S})$ where:

- $(\mathcal{X}, d)$ is a complete metric space. We shall call it the *iterative space*.
- $\mathcal{S}$ is a semigroup acting on points of $\mathcal{X}$ such that : $p \mapsto T \circ p$, where $T \in \mathcal{S}$ is a contractive operator. $\mathcal{S}$ is called the *iterative semigroup*.

The space $\mathcal{H}(\mathcal{X})$ denoting the set of non-empty compacts of $\mathcal{X}$ is a complete metric space, endowed with the HAUSDORFF distance:

$$d_H(K, K') = \max\{\max_{p \in K} \min_{q \in K'} d(p, q), \max_{q \in K'} \min_{p \in K} d(p, q)\}.$$

An IFS is a finite subset of the semigroup $\mathcal{S}$:

$$\mathcal{T} = \{T_1, ..., T_N\}.$$

The associated HUTCHINSON operator is:

$$K \in \mathcal{H}(\mathcal{X}) \mapsto \mathcal{T} \circ K = T_1 \circ K \cup ... \cup T_N \circ K.$$

This operator is contractive in the metric space $(\mathcal{H}(\mathcal{X}), d_H)$ and thus admits a fixed point $\mathcal{A}(\mathcal{T})$ given by:

$$\mathcal{A}(\mathcal{T}) = \lim_{k \to \infty} \mathcal{T}^k \circ K_0,$$

for any non empty compact set K_0.

IFS attractors constitute a particular family of compacts. They are *invariant sets* for $\mathcal{T}$:

$$\mathcal{A} = T_1 \circ \mathcal{A} \cup ... \cup T_N \circ \mathcal{A}, \text{ with } T_i \in \mathcal{T}.$$

Example 1. We present on Figure 1 an IFS given in [PH91]. By convention,

- $T(a, b, c)$ denotes the translation by the vector (a, b, c).
- $Rx(a)$ denotes the rotation of angle a around the Ox axis.
- $H(a)$ denotes the scaling of ratio a with respect to the origin of the coordinate system.

$\mathcal{T}_{tree} = \{T_1, T_2, T_3, T_4\}$

$T_1 = H(0.5);$
$T_2 = T(0, 0.5, 0) \circ H(0.5);$
$T_3 = T(0, 1, 0) \circ Rx(\pi/4) \circ H(0.5);$
$T_4 = T(0, 1, 0) \circ Rx(-\pi/4) \circ H(0.5).$

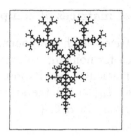

Fig. 1: Example of IFS.

IFS techniques are an interesting approach for modeling fractals. In addition to the simplicity of their mathematical formalism, they provide an efficient tool to encode complex shapes with a small set of contractive operators. This paper focuses on affine operators.

However, IFS techniques suffer from important drawbacks when used for geometric modeling purposes. These consist of two essential restrictions. Firstly, the attractors are strictly self-affine. Secondly, there is no convenient way to manipulate these attractors. Such a deficiency is an important limit of fractal geometric modeling.

2.2 Generalizations of IFS techniques

Many authors have proposed generalizations of the IFS model. These generally consist in constructing subsets of an IFS attractor:

$$K_\omega \subseteq \mathcal{A}(\mathcal{T}) \text{ with } K_\omega = \lim_{n \to \infty} K_n \text{ and } K_n \subseteq \mathcal{T}^n \circ K_0.$$

Two different approaches can be distinguished. Firstly, a "discrete" approach based on quadtrees or octrees [BM89] [CD93]: $\mathcal{A}(\mathcal{T}) = [0, 1]^2$ or $[0, 1]^3$. Secondly, a "continuous" approach in which the IFS is arbitrary [Gen92] [PH92] [TT93] [DTG95]. The discrete approach is often used for image compression.

Moreover, in each case there are different techniques to define the restriction of the IFS, such as formal languages, equations systems, trees, graphs, *etc*. These approaches are not totally equivalent. That's why we have looked for a model that could integrate a wide part of these models. Moreover we needed a model on which operations could easily be defined.

A generalization based on a matricial formalism is proposed in this paper. This approach integrates not only the existing models based on equations systems or graphs but also those based on regular languages (finite automaton, grammars). Moreover, a set of operations is given for a constructive approach of fractal attractors using basic operations such as union and intersection.

3 IFS matrices

Our approach is based on the HUTCHINSON operator, generalized by PEITGEN, JÜRGENS and SAUPE [PJS92].

3.1 Space of matrices of compacts

We define matrices which components are compact sets. As the HAUSDORFF distance is defined for non empty compacts, we should generalize the HAUSDORFF distance in a particular way in order to allow empty components.

We denote $M_R(\mathcal{H}(\mathcal{X}))$ the set of matrices in $\mathcal{H}(\mathcal{X})$ where the non empty elements are given by $R \subseteq \{1, \ldots n\} \times \{1, \ldots n\}$. That is to say for all matrix $\tilde{K} \in M_R(\mathcal{H}(\mathcal{X}))$,

$$\tilde{K}_{ij} \in \mathcal{H}(\mathcal{X}) \cup \emptyset,$$
$$\tilde{K}_{ij} \neq \emptyset \Leftrightarrow (i,j) \in R.$$

We can now define a distance on $M_R(\mathcal{H}(\mathcal{X}))$: let $\tilde{K}$ and $\tilde{K}'$ be in $M_R(\mathcal{H}(\mathcal{X}))$,

$$d_\infty(\tilde{K}, \tilde{K}') = \max_{(i,j) \in R} d_H(\tilde{K}_{ij}, \tilde{K}'_{ij}).$$

As R is a finite set and $(\mathcal{H}(\mathcal{X}), d_H)$ a metric space, $(M_R(\mathcal{H}(\mathcal{X})), d_\infty)$ is a complete metric space.

3.2 IFS Matrix

We now define a matrix which components are IFS. Let $\mathcal{S}$ be an affine semigroup. We define a matrix of IFS as an $n \times n$ matrix which components are finite, possibly empty, subsets of $\mathcal{S}$. We denote $\mathcal{P}_+(\mathcal{S})$ the set of finite subsets of $\mathcal{S}$ and $M_R(\mathcal{P}_+(\mathcal{S}))$ the set of matrices of IFS.

Let $\tilde{H} = (\tilde{H}_{ij}) \in M_R(\mathcal{P}_+(\mathcal{S}))$

$$\tilde{H}_{ij} = \begin{cases} \bigcup_{k \in I_{ij}} T_k & \text{if } (i,j) \in R \\ \emptyset & \text{otherwise} \end{cases},$$

with $T_k \in \mathcal{S}$ and $I_{ij} \subseteq [1 \ldots N]$ where $N = Card(\bigcup_{ij} \tilde{H}_{ij})$.

Example 2. Let $\mathcal{T}_{tree}$ be the IFS previously defined. We can choose the following matrix of IFS:

$$\tilde{H}_{tree} = \begin{pmatrix} \{T_3, T_4\} & \{T_1, T_2\} \\ \emptyset & \{T_1, T_2\} \end{pmatrix}.$$

$\tilde{H}_{tree} \in M_R(\mathcal{P}_+(\mathcal{S}))$ with $R = \{(1,1), (1,2), (2,2)\}$.

We can now define the composition of these matrices.
Let $\tilde{H} \in M_R(\mathcal{P}_+(\mathcal{S}))$, $\tilde{K} \in M_S(\mathcal{H}(\mathcal{X}))$ and $\tilde{H}' \in M_S(\mathcal{P}_+(\mathcal{S}))$:

$$(\tilde{H} \circ \tilde{K})_{ij} = \bigcup_{k=1..n, (i,k) \in R, (k,j) \in S} \tilde{H}_{ik} \circ \tilde{K}_{kj} \in M_{R \circ S}(\mathcal{H}(\mathcal{X})).$$

$$(\tilde{H} \circ \tilde{H}')_{ij} = \bigcup_{k=1..n, (i,k) \in R, (k,j) \in S} \tilde{H}_{ik} \circ \tilde{H}'_{kj} \in M_{R \circ S}(\mathcal{P}_+(\mathcal{S})).$$

with $R \circ S = \bigcup_{(i,k) \in R, (k,j) \in S} \{(i,j)\}$.

Moreover if the relation R is a transitive one, then $M_R(\mathcal{P}_+(\mathcal{S}))$ is stable by composition, that is:

$$\tilde{H} \in M_R(\mathcal{P}_+(\mathcal{S})), \tilde{K} \in M_R(\mathcal{H}(\mathcal{X})) \Rightarrow \tilde{H} \circ \tilde{K} \in M_R(\mathcal{H}(\mathcal{X}))$$

and $\tilde{H}$ is a contractive operator on $(M_R(\mathcal{H}(\mathcal{X})), d_\infty)$.

3.3 Attractor matrix

We now define an attractor matrix as the limit of an iterated matrix of IFS. Such a limit exists when the matrix of IFS is aperiodic. Intuitively, this means that empty components of the matrix will be stable during the iteration. This hypothesis is necessary because of the HAUSDORFF distance.

More precisely, we define $\tilde{H} \in M_R(\mathcal{P}_+(\mathcal{S}))$ as an aperiodic matrix when it exists $m \in \mathbb{N}$ such that

$$\forall k > m, R^m = R^k.$$

Let $S = R^m$, then $M_S(\mathcal{P}_+(\mathcal{S}))$ is stable by $\circ$ and

$$\forall k > m, \tilde{H}^k \in M_S(\mathcal{P}_+(\mathcal{S})).$$

Moreover, if $\tilde{K}$ is a diagonal matrix, we have

$$\forall k > m, \tilde{H}^k \circ \tilde{K} \in M_S(\mathcal{H}(\mathcal{X})).$$

We can now define an attractor matrix in the same way as the attractor of an IFS. Indeed, let $\tilde{H}$ be an aperiodic matrix, as $\tilde{H}$ is a contractive operator, we can prove that the sequence $(\tilde{H}^k \circ \tilde{K})_{k>m}$ is a CAUCHY sequence in the complete metric space $(M_S(\mathcal{H}(\mathcal{X})), d_\infty)$.

The limit of such a sequence is thus defined, and we call it the attractor matrix:

$$\tilde{A}(\tilde{H}) = \lim_{n \to \infty} \tilde{H}^n \circ \tilde{K}.$$

Denoting $\mathcal{T} = \bigcup_{(i,j) \in R} \tilde{H}_{ij}$, we have

$$\tilde{A}(\tilde{H})_{ij} \subseteq \mathcal{A}(\mathcal{T}).$$

Example 3. In the case of the IFS $\mathcal{T}_{tree} = \{T_1, T_2, T_3, T_4\}$, with the matrix of IFS:

$$\tilde{H}_{tree} = \begin{pmatrix} \{T_3, T_4\} & \{T_1, T_2\} \\ \emptyset & \{T_1, T_2\} \end{pmatrix},$$

we have

$$\tilde{H}^k_{tree} = \begin{pmatrix} \{T_3, T_4\}^k & \bigcup_{i+j=k-1} \{T_3, T_4\}^i \{T_1, T_2\}^j \\ \emptyset & \{T_1, T_2\}^k \end{pmatrix},$$

thus

$$\tilde{A}(\tilde{H}_{tree}) = \lim_{n \to \infty} \tilde{H}^n \circ \tilde{K} = \begin{pmatrix} \mathcal{A}(T_3, T_4) & \mathcal{A}(T_3, T_4) \cup \{T_3, T_4\}^* \circ \mathcal{A}(T_1, T_2) \\ \emptyset & \mathcal{A}(T_1, T_2) \end{pmatrix}.$$

$\tilde{A}(\tilde{H}_{tree})$ is shown Figure 2.

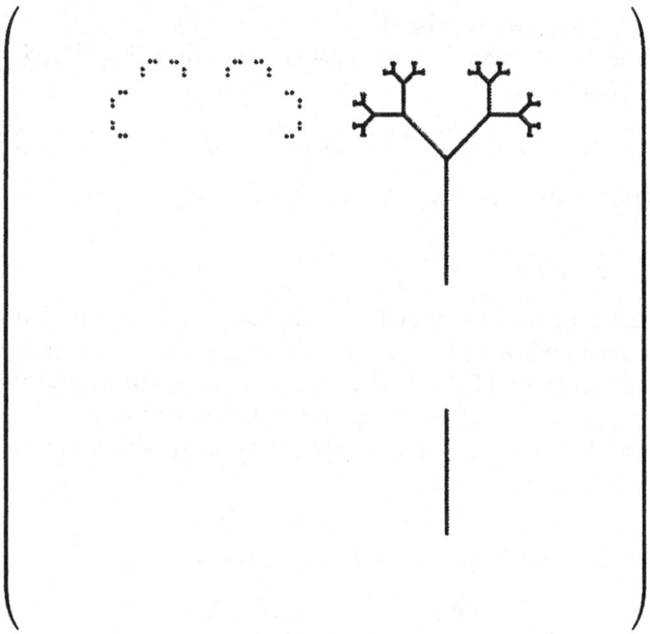

Fig. 2: Attractor matrix $\tilde{\mathcal{A}}(\tilde{H}_{tree})$.

We have proved the equivalence of our model with the existing techniques, in particular we have worked on the relation between our model and formal languages ones [Tho96].

In the case of a model based on finite automata, the idea of this relation is to select some of the components of the matrix. The choice is made according to the initial and final states of the automaton. The attractor associated to the automata is then the union of the chosen components.

4 Set manipulations

The matrix model permits combination of two IFS matrices to obtain a new attractor matrix. We have studied two types of operations, some general operations on matrices, such as union, composition, intersection, tensor product [Tho96]; and some operations associated with formal language theory, such as tensor sum, block matrix [TT95].

These operations provide us with a constructive approach. Indeed, it is possible to combine two IFS matrices and to obtain a resultant matrix that we combine with another one, and so on. Before illustrating this approach with an example, let us give a brief definition of each operation.

4.1 Matrices operations

Let $\tilde{H}$ and $\tilde{H}'$ be two matrices of IFS of size respectively n and m.

The union and intersection operations are defined in the following way:

$$(\tilde{H} \cup \tilde{H}')_{ij} = \tilde{H}_{ij} \cup \tilde{H}'_{ij},$$

$$(\tilde{H} \cap \tilde{H}')_{ij} = \tilde{H}_{ij} \cap \tilde{H}'_{ij}.$$

Composition has already been defined:

$$(\tilde{H} \circ \tilde{H}')_{ij} = \bigcup_k \tilde{H}_{ik} \circ \tilde{H}'_{kj}.$$

Tensor product is an external operation:

$$\tilde{H}'' = \tilde{H} \otimes_\cup \tilde{H}' = \begin{pmatrix} \tilde{H}_{11} \cup \tilde{H}' \dots \tilde{H}_{1n} \cup \tilde{H}' \\ \vdots \qquad\qquad \vdots \\ \tilde{H}_{n1} \cup \tilde{H}' \dots \tilde{H}_{nn} \cup \tilde{H}' \end{pmatrix},$$

with $\tilde{H}''$ a square matrix of size $n.m$, and

$$\tilde{H}_{ij} \cup \tilde{H}' = \begin{pmatrix} \tilde{H}_{ij} \cup \tilde{H}'_{11} \dots \tilde{H}_{ij} \cup \tilde{H}'_{1m} \\ \vdots \qquad\qquad \vdots \\ \tilde{H}_{ij} \cup \tilde{H}'_{m1} \dots \tilde{H}_{ij} \cup \tilde{H}'_{mm} \end{pmatrix}.$$

4.2 Language operations

Tensor summation is an operation associated with language theory. This operation corresponds to the shuffle (denoted by $\sqcup$) of two regular languages. For this reason, we denote it "$\Diamond_\sqcup$". We define it by

$$\tilde{H}'' = \tilde{H} \Diamond_\sqcup \tilde{H}' = \begin{pmatrix} \tilde{Id}.\tilde{H}_{11} \cup \tilde{H}' & \tilde{Id}.\tilde{H}_{12} & \dots & \tilde{Id}.\tilde{H}_{1n} \\ \tilde{Id}.\tilde{H}_{21} & \tilde{Id}.\tilde{H}_{22} \cup \tilde{H}' & \dots & \tilde{Id}.\tilde{H}_{2n} \\ \vdots & & & \vdots \\ \tilde{Id}.\tilde{H}_{n1} & \tilde{Id}.\tilde{H}_{n2} & \dots & \tilde{Id}.\tilde{H}_{nn} \cup \tilde{H}' \end{pmatrix},$$

with $\tilde{H}''$ a square matrix of size $n.m$ and $\tilde{Id}$ the identity matrix of size m.

The block matrix operation is a simple combination of two matrices:

$$\tilde{H}'' = \tilde{H} \Diamond_\sqcup \tilde{H}' = \begin{pmatrix} \tilde{H} & \emptyset \\ \emptyset & \tilde{H}' \end{pmatrix}.$$

As for the tensor summation, this operation has been inspired by language theory. It corresponds to the union of two regular languages. Using this operation we can define the union of two fractal shapes.

Moreover, if we replace some $\emptyset$ with the identity in the resulting matrix, we obtain the concatenation of two regular languages. We will denote this operation by "$\Diamond$".

4.3 Constructive approach

We show Figure 3 some attractor matrices made for a project "art and science". These pictures have been made by Martine Rondet-Mignotte. She is an artist and uses our work to produce artistic images. This figure shows how some complex shapes can be constructed, using the "step by step" approach, with the use of the matrix operations.

Once the resulting matrix has been constructed, one can choose some of the components. This allows to construct an object which is the union of some components. Figure 4 shows an example of objects extracted from the resulting matrix of Figure 3

Figure 5 shows some other results of operations.

4.4 Classification by inclusion

To facilitate the choice of operations, we have ordered (in the sense of set inclusion) the results of these operations. We consider the compact union of all components of the attractor matrix:

$$\mathcal{A}(\tilde{H}) = \bigcup_{i,j \in \{1,\dots n\}} \tilde{\mathcal{A}}(\tilde{H})_{ij}.$$

Denoting by $\mathcal{T}$ the IFS composed of all the transformations used in both IFS matrices $\tilde{H}_1$ and $\tilde{H}_2$, we can prove the following inclusions [Tho96], where $\downarrow$ denotes inclusion.

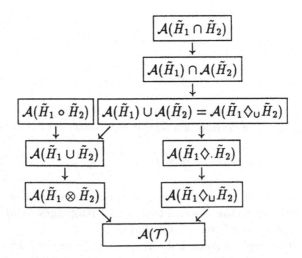

Figure 6 shows an example of inclusion.

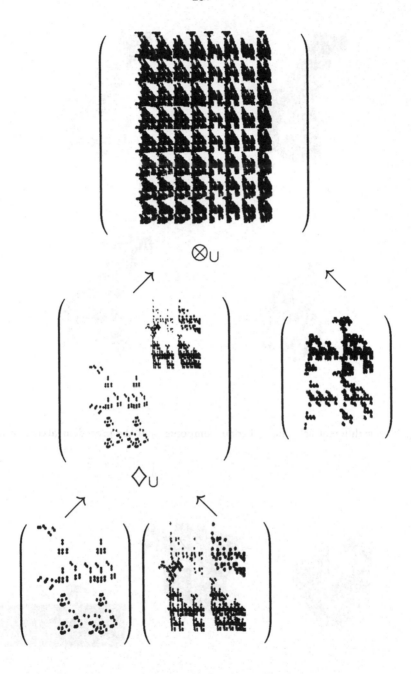

Fig. 3: Constructive tree.

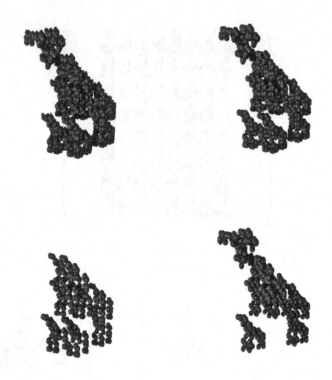

Fig. 4: Four different selections of some components of the precedent attractor matrix.

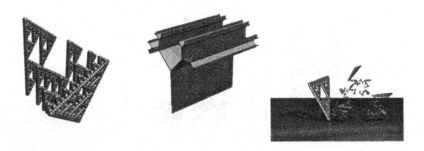

Fig. 5: Examples of 3D attractors

233

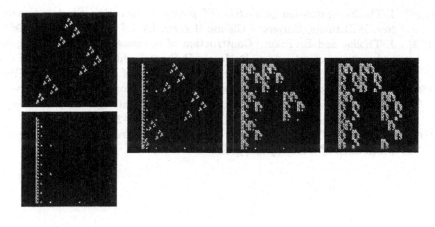

$\mathcal{A}(\tilde{H}_1)$ and $\mathcal{A}(\tilde{H}_2)$ $\quad \subset \mathcal{A}(\tilde{H}_1 \Diamond_\cup \tilde{H}_2)$ $\quad \subset \mathcal{A}(\tilde{H}_1 \cup \tilde{H}_2)$ $\quad \subset \mathcal{A}(\tilde{H}_1 \otimes_\cup \tilde{H}_2)$

Fig. 6: Example of inclusion.

5 Conclusion

Using a generalization of the classical IFS model, we construct a fractal model that enables a constructive approach.

This approach is a part of a project of fractal modeler. In this project, we have also worked on free forms techniques applied to fractals [ZT96]. We should now combine both approaches to obtain a complete system of fractal modeling.

References

[Bar88] M.F. Barnsley. *Fractals Everywhere*. Academic press, INC, 1988.

[BM89] J. Berstel and M. Morcrette. Compact representation of patterns by finite automata. In *Proceedings of Pixim*, pages 387–395, 1989.

[CD93] K. Culik II and S. Dube. Rationnal and affine expressions for image synthesis. *Discrete Appl. Math.*, 41:85–120, 1993.

[DTG95] S. Duval, M. Tagine, and D. Ghazanfarpour. Modélisation de fractals par les arbres étiquetés. In *3emes journées AFIG, Marseille*, pages 125–134, novembre 1995.

[Gen92] C. Gentil. *Les Fractales en Synthèse d'Images : le Modèle IFS*. PhD thesis, Université Claude Bernard LYON 1, France, 1992.

[PH91] P. Prusinkiewicz and M.S. Hammel. Automata, languages and iterated function systems. In *lecture notes for the SIGGRAPH'91 course : "Fractal modeling in 3D computer graphics and imagery"*, 1991.

[PH92] P. Prusinkiewicz and M.S. Hammel. Escape-time visualization method for language-restricted iterated function systems. In *Proceedings of Graphics Interface'92*, May 1992.

[PJS92] H.O. Peitgen, H. Jürgens, and D. Saupe. Encoding images by simple transformations. In *Fractals For The Classroom, New York*, 1992.

[Tho96] J. Thollot. *Extension du Modèle IFS pour une Géométrie Fractale Constructive*. PhD thesis, Université Claude Bernard LYON 1, France, sept 1996.

[TT93] J. Thollot and E. Tosan. Construction of fractales using formal languages and matrices of attractors. In Harold P. Santos, editor, *Proceedings of Compugraphics'93*, pages 74–81, Technical University of Lisbon, december 1993.

[TT95] J. Thollot and E. Tosan. Constructive fractal geometry : constructive approach to fractal modeling using languages operations. In *Proceedings of Graphics Interface'95, Quebec, Canada*, pages 196–203, may 1995.

[ZT96] C. Zair and E. Tosan. Fractal modeling using free forms techniques. In *Proceedings of Eurographics*, 1996.

Ray-Tracing and 3-D Objects Representation in the BCC and FCC Grids

Luis Ibáñez, Chafiaâ Hamitouche, Christian Roux

Département Image et Traitement de l'Information
ENST - Bretagne, Technopôle de Brest-Iroise
BP 832, 29285 Brest, France
E-mail: Luis.Ibanez@enst-bretagne.fr

Abstract. This paper describes a ray-tracing and an object description method for objects sampled not in the usual cubic grid, but in BCC (Body Centered Cubic) and FCC (Face Centered Cubic) grids, which are well known in crystallography. The use of this kind of grids is motivated by their interesting characteristics: they reduce the density of samples needed to represent a signal without information loss and they have better topological properties than the cubic grid.

keywords: Image Representation, Rendering, Visualization

1 Introduction

Sampling a signal in $3 - D$ space satisfying the Shannon theorem leads to the use of the BCC grid as the optimal sampling grid [1]. Its reciprocal, the FCC grid, shows also a good performance from the point of view of low sample density. These grids have more symmetries than the cubic one, which gives a better approach to an anisotropic data representation. For this reason they are recommended as optimal grids for Mathematical Morphology processing [3]. The use of these grids simplifies the surface extraction process in volume data [14]. Recently it has been shown that the use of the BCC and FCC grids enhances the performance of reconstruction from projections methods [10], which leads to the emergence of volume data in BCC and FCC format in 3-D imaging for medical applications.

In spite of their interesting characteristics, the use of these grids is usually rejected because of the increment in algorithmic complexity. This argument is not always true, and it reflects only the lack of deep work in the analysis of the discrete topology corresponding to these grids. The method presented here shows that it is possible to exploit the potential offered by the BCC and FCC grids.

2 Ray Tracing in Volume Data

Direct Volume Rendering methods include a ray tracing process in order to compute the contribution of the voxels in the volume to each pixel in the synthesized

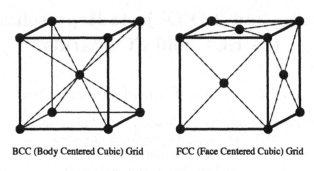

BCC (Body Centered Cubic) Grid   FCC (Face Centered Cubic) Grid

Fig. 1. BCC and FCC grids

image[9][6][5][7]. When objects are represented by their discrete surfaces, topological problems arise in the cubic grid, which leads to the possibility of passing a ray through a surface without touching it [4]. To avoid this problem, thicker surfaces or thicker rays should be used. However thicker rays implies a longer processing time in the ray tracing process.

BCC and FCC grids do not lead to such topological problems, and allows in a more robust way the notions of discrete curves, surfaces and objects to be defined. Here, the objects are represented on a BCC or FCC grid by using only their external surfaces. A discrete straight line tracing algorithm has been developed to guide the ray tracing process from each pixel towards the objects in the volume data. Interactions between the ray and the objects surfaces are defined in a discrete way too.

3 Topology of the BCC and FCC grids

Topology is defined here in function of the Voronoi cells of each point in the grid. The $3-D$ shape associated to the voxel is the solid polyhedron corresponding to the Voronoi cell (see Figure 3). A connected set of voxels is considered as a $3-D$ Manifold. The Voronoi cells can fill the space by a simple tiling, and the volume data can be considered here as region of the $3-D$ space completely filled with polyhedral voxels.

Definition 1. Two n-dimensional regions of the space are *connected* if they have a $(n-1)$-dimensional region in common. Two voxels are connected if they have a common face. Two faces are connected if they have a common edge. Two edges are connected if they have a common vertex. The existence of m-dimensional regions ($m < (n-1)$) in common between two n-dimensional regions does not implies connectivity.

Definition 2. An *Object* is a set of connected voxels. An *Object Surface* is the set of faces between object voxels and non-object voxels.

The Voronoi cell in the BCC grid is the Truncated Octahedron, and in the FCC grid is the Rhombic Dodecahedron [2]. BCC and FCC grids have the nice property that an edge in the tiling structure is common to exactly three voxels, and hence are common to exactly three faces. When these voxels are classified as belonging to an object or not, always two out of the three voxels belongs to the same class. Then two out of the three faces belong to the object surface. This local property implies that every face in the object surface has exactly one face connected to each of its edges. Then object surfaces are always closed surfaces.

4 Object Representation

Volume data represents a $3-D$ signal which has been discretized to be processed. Often the original source of the signal is a continuous real process (as is the case in medical imaging). A classification process should be done in order to determine the regions corresponding to an object. Classification is hard to perform close to the object boundary regions, which leads to a higher classification error for such voxels and therefore to errors in the shape of the external object surface.

When the description of the object surface is limited to the determination of faces belonging or not to the surface, certain information about the real form of the object boundary is lost, and hence the original form cannot be retrieved exactly. This is specially important for processes that must use the surface normal. To avoid this problem a certain amount of information should be left from the classification process to indicate the relative position of a voxel in the transition region.

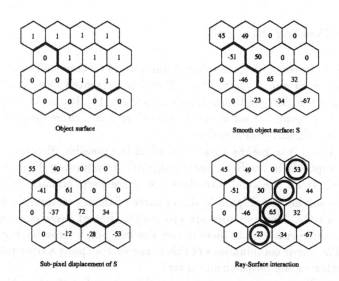

Fig. 2. Illustration of smooth curve representation

Definition 3. A *Smooth Object Surface* is the set of all the voxels in contact with a face of the *Object Surface*, along with a signed integer coefficient assigned to each voxel. The coefficient represents the relative position of the voxel in the transition region. This coefficients take values from -128 to 127.

Objects represented in this way allow the information of the complete volume data to be reduced to the representation of object surfaces that it contains, without loosing their original smoothness. Several advantages are found in this representation: objects can be deformed or displaced by sub-voxel distances only by changing the coefficients (displacements greater than the voxel size will change also the voxels which belong to the smooth surface). Interactions between object surfaces can be performed in a more realistic way, (collisions, soft contacts). Interactions between object surfaces and light beams can avoid aliasing effects originated by the limited set of orientations of the voxels faces.

Figure 2. Illustrates the object, the faces in its surface (top left). The smooth surface representation with coefficients that are positive on one side, and negatives on the other (top right). The smooth surface after a sub-voxel displacement (bottom left). The interaction between a ray (voxels with circles) and the smooth surface (bottom right). Surface normals are easily evaluated by using the coefficients of the voxels in the smooth surface representation.

To obtain the coefficients for the surface representation, a signal processing method is used. An edge detector operator based on geometric moments is applied to estimate the sub-voxel distance between the voxel center and the object boundary [12][13]. It retrieves positive values for the interior voxels and negative values for the exterior voxels.

5 Line Tracing Method

To cast a light beam, a discrete straight line should be traced. The line is defined by an origin point and a vector. The algorithm presented here is as simple and performant as the classical Bresenham's algorithm [11]. It should be noted that a light beam is a connected set of voxels, therefore it is a three dimensional object and not a one-dimensional manifold.

The first step is to realize a partition of all the possible directions in the grid (which corresponds, in the Bresenham's algorithm, to finding the octant in which lies the line). Here, the valid directions are those perpendicular to the faces of the Voronoi polyhedron. Figure 3, shows three of them (vectors $\vec{A}$, $\vec{B}$, $\vec{C}$ in the two Voronoi cells. These vectors are grouped in sets of three elements. Vectors in the same set correspond to faces of the Voronoi polyhedron having a vertex in common. The linear combinations of this three vector spans a triangular pyramid with its vertex in the polyhedron center.

Figure 3. Illustrate the geometric configuration for the two Voronoi cells. Vector $\left\{ \vec{A}, \vec{B}, \vec{C} \right\}$ are normal to the voxel faces with the common vertex V. The region defined by the points $\{a, e, b, f, c, d\}$ is the intersection of the pyramid spanned by vectors $\left\{ \vec{A}, \vec{B}, \vec{C} \right\}$ with the polyhedron surface. Any grid vector originated in the polyhedron center and passing through this region can be ex-

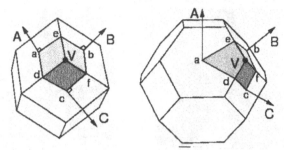

Rhombic Dodecahedron (FCC)   Truncated Octahedron (BCC)

Fig. 3. Partition of directions

pressed as a linear combination the vectors $\left\{\vec{A}, \vec{B}, \vec{C}\right\}$ with integer positive coefficients. There are 24 of these regions in the BCC, and 20 in the FCC grid.

The components $\{n_a, n_b, n_c\}$ of the line vector PQ in the basis $\left\{\vec{A}, \vec{B}, \vec{C}\right\}$, means geometrically that to go from P to Q, n_a steps must be performed in A direction, n_b in B direction and n_c in C direction. Figure 4 (left), represents the line tracing process from point P to point Q. N is the plane normal to the line. R is the rhombic base of the prism which contains all the line beam points. The algorithm minimize the area of the rhombus R. Only integer sums should be computed to advance from one point to the next. Figure 4 (right) shows the projection of the grid points on the plane orthogonal to PQ, it is composed of a set of $2 - D$ grids shifted by a constant vector.

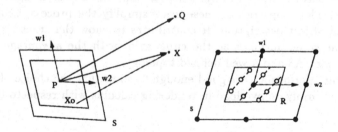

Fig. 4. Ray casting scheme

6 Results

The object representation method, and the ray casting algorithm described here, is used as a ray casting engine in a direct volume rendering program. The BCC grid is used as support of the volume data. Figure 5 shows the rendering of

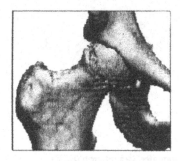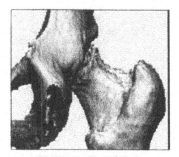

Fig. 5. Medical data volume in BCC grid support (Hip Joint)

a medical image using this program. The algorithm is implemented in C++, on a PC-Pentium 166MHz platform, running Linux 2.0.18. The original volume corresponds to a CT reconstruction of the hip joint containing 100x100x100 voxels. It has been resampled using a 3D cubic B-Spline interpolator, to obtain a BCC structure. This resampling process duplicate the number of voxels. Output image has 256x256 pixels. Rendering time is 3s, it includes ray tracing and lighting calculus with one light source, with antialiasing.

7 Conclusions

The methods described in this article performs ray casting in volume data sampled with the BCC and FCC grids. The object representation is complete enough to allow process as object to object interaction, and object to light interaction to be done. The properties of these grids simplify the process involved in ray tracing and object description. It contributes to show that these grids can be used for the same purposes as the cubic grid, with the advantage of reduced number of samples and a well defined topology.

Algorithm performance is good enough to show that the choice of this topology does not imply an overload in rendering calculus with respecto to the cubic grid.

References

1. D.E.Dudgeon, R.M. Mersereau *Multidimensional digital signal processing* Prentice-Hall, Englwood Cliffs; NJ; 1984.
2. C.Kittel *Introduction to solid-state physics* John Wiley & Sons Inc. 1971.
3. J.Serra *Image Analysis and Mathematical Morphology* . Academic Press Inc. 1982.
4. Arie E. Kaufman, *Volume Synthesis* 6th International Workshop, Discrete Geometry for Computer Imagery 96, Lyon, France, November 1996. Lecture Notes in Computer Science, Springer Verlag.

5. J.J.Jacq, C.Roux, *A Direct Multi-Volume Rendering Method Aiming at Comparisions of 3-D Images and Models* IEEE Transactions on information technology in biomedicine, Vol 1., No.1, pp 30-43, march 1997.

6. G.T.Herman, *3D Display: A Survey From Theory to Applications* Computerized Medical Imaging and Graphics, Vol. 17, Nos 4/5, pp 231-242, 1993.

7. W.Krüger, P.Schröder. *Data parallel volume rendering* pp 37-52. in Scientific Visualization, Academic Press, 1994.

8. U.Tiede, et al, *Investigation of Medical 3D-Rendering Algorithms* IEEE Computer Graphics & Applications, Vol. 10, No.2, pp 41-53, 1990.

9. G.Sakas, M.Grimm, A.Savapoulos *Optimized Maximum Intensity Projection (MIP).* in Rendering Techniques'95, Proceedings of the Eurographics Workshop in Dublin, Ireland, June 12-14, Springer Verlag, 1995.

10. S. Matej, R.M. Lewit *Efficient 3-D Grids for Image Reconstruction Using Spherically Symmetric Volume Elements* IEEE Transactions on Nuclear Science, Vol 42, No.4, August 1995, pp 1361-1370.

11. J.E. Bresenham *Algorithm for computer control of a digital plotter* IBM Systems Journal, 1965, Vol 4. pp 25-30.

12. Li Min Luo, Ch. Hamitouche, L. Dillenseger, J.L. Coatrieux *A Moment-Based Three-Dimensional Edge Operator*, IEEE Transactions on Biomedical Engineering,. Vol 40, No.7, Jul 1993. pp 693-703.

13. L. Ibáñez, C. Hamitouche, C. Roux *Moment-based operator for sub-voxel surface extraction in medical imaging* International Conference on Image Processing ICIP'96, Lausanne, Switzerland, September 1996.

14. L. Ibáñez, C. Hamitouche, C. Roux *Determination of discrete sampling grids with optimal topological and spectral properties* 6th International Workshop, Discrete Geometry for Computer Imagery 96, Lyon, France, November 1996. Lecture Notes in Computer Science, Springer Verlag.

Supercover of Straight Lines, Planes and Triangles

Eric Andres (*), Philippe Nehlig (**), and Jean Françon (**)

(*) TUCS - Turku Centre for Computer Science
Lemminkäisenkatu 14A, 20520 Turku, Finland
(**) Laboratoire des Sciences de l'Image, de l'Informatique et de la Télédétection
Université Louis Pasteur
7 rue René Descartes, 67084 Strasbourg Cedex, France
`{andres, nehlig, francon}@dpt-info.u-strasbg.fr`

Abstract
The Supercover of a Euclidean object is the set of the pixels or voxels intersected by the object. The Supercover of 2D lines and 2D triangles are defined analytically. Some geometric properties, localization, and generation algorithms are given. The same is done for 3D lines, planes, and 3D triangles.

Keywords : Discrete 3D Modelling, Discrete Lines, Discrete Planes, Discrete Polygons, Supercover.

1. Introduction

The *Supercover* of a continuous (Euclidean) object is the set of the pixels or voxels intersected by the object, where a pixel (resp. voxel) is a unit square (resp. cube) of $\mathbf{R}^2$ (resp. $\mathbf{R}^3$) centered on a point of integer coordinates (integer point) of $\mathbf{Z}^2$ (resp. $\mathbf{Z}^3$), the discrete plane (resp. space). This concept of discretization is not new. Recently, Cohen and Kaufman [Cohe95] reactivated this concept for volume graphics purposes. In fact, supercovers have interesting properties of tunnel-freeness, and interesting properties under set operations. More recently, [ASA96] introduces for the first time discrete analytical descriptions of the Supercover of a 2D line defined by two rational points, and of a 3D plane defined by three non-colinear rational points. Thus, discrete analytical modelling becomes possible.

While in [ANF97] we extended [ASA96] by introducing discrete analytical descriptions of the Supercover of a 3D line, of a polygon, and of a polyhedron for discrete geometric modelling purposes, giving their generation algorithms, in the present paper, we develop systematically the geometric properties of the Supercover of a 2D line, of a 2D triangle, of a 3D line, of a 3D plane, and of a 3D triangle. We develop also the basic generation algorithms as well as point localizing algorithms. The complexity of our algorithms, in the general case, is proportional to the number of generated points, and can be improved using usual optimisation techniques.

The paper is organised as follows. In section 2 we present the notations and the main definitions used in this paper. In section 3.1 we give formula for the Supercover of points and of boxes; in section 3.i, for i=2,3. we deal with the definition, properties and algorithms of the supercovers of primitives of dimension i. We conclude in section 4.

The algorithms are elementary consequences of the main theorems. The main theorems are the characterization of the Supercover of :

(1) a 2D straight line (resp. 3D plane) by an arithmetic line (resp. plane);
(2) a 3D straight line by the Supercover of its 3 projections on the orthotropic planes;
(3) a 2D or a 3D triangle by set operations on the preceding supercovers.

The proofs are either very elementary or long, technical, and without giving more light on the asserted propositions. We give thus only sketches for some of them.

2. Preliminaries

The set of the real numbers is noted $\mathbf{R}$, the set of the rational numbers $\mathbf{Q}$, the set of the integers $\mathbf{Z}$, and the set of the strictly positive integers $\mathbf{N}^*$. $\lfloor x \rfloor$ is the greatest integer less than or equal to x. $\lceil x \rceil$ is the least integer greater than or equal to x. If not specified differently, $\left\{ \frac{p}{q} \right\}$, where $\frac{p}{q}$ is rational, is a Euclidean remainder: $\left\{ \frac{p}{q} \right\} = p - q \left\lfloor \frac{p}{q} \right\rfloor$. The greatest common divisor of the integers $a_1,...,a_n$ is noted $gcd(a_1,...,a_n)$. The dimension of the space is noted n. In this article we limit ourselves to dimension 1 to 3. A discrete point is a point in $\mathbf{Z}^n$. A discrete object is a set of discrete points. A Euclidean, also called continuous, object is a set of points in $\mathbf{R}^n$. For a discrete object A, $\overline{A} = \mathbf{Z}^n \backslash A$.

A *pixel* $V(X)$, $X(x)$ a 1D discrete point, is defined by the Euclidean interval $\left[x - \frac{1}{2}, x + \frac{1}{2} \right]$. A pixel $V(X)$, $X(x,y)$ a 2D discrete point, is defined by the Euclidean square $\left[x - \frac{1}{2}, x + \frac{1}{2} \right] \times \left[y - \frac{1}{2}, y + \frac{1}{2} \right]$. A *voxel* $V(X)$, $X(x,y,z)$ a 3D discrete point, is defined by the Euclidean cube $\left[x - \frac{1}{2}, x + \frac{1}{2} \right] \times \left[y - \frac{1}{2}, y + \frac{1}{2} \right] \times \left[z - \frac{1}{2}, z + \frac{1}{2} \right]$. Each pixel (voxel) has a unique corresponding discrete point and vice-versa.

The *Supercover* S^ω of a Euclidean object S is the set of all the discrete points X with corresponding pixel (voxel) $V(X)$ such that $V(X) \cap S \neq \varnothing$.

For a 2D point $A(x,y)$, $A_x=(x)$ and $A_y=(y)$ are 1D points resulting of the orthogonal projections of A on Ox and Oy. For a 3D point $A(x,y,z)$, $A_x=(x)$, $A_y=(y)$ $A_z=(z)$ are 1D points resulting of the orthogonal projections of A on Ox, Oy and Oz respectively. For a 3D point $A(x,y,z)$, $A_{xy}=(x,y)$, $A_{xz}=(x,z)$ and $A_{yz}=(y,z)$ are 2D points resulting of the orthogonal projections of A on Oxy, Oxz and Oyz respectively.

Let T be a 2D discrete object, then $T^x = \{(x,y,z) \mid (y,z) \in T \text{ and } x \in Z\}$; $T^y = \{(x,y,z) \mid (x,z) \in T \text{ and } y \in Z\}$ and $T^z = \{(x,y,z) \mid (x,y) \in T \text{ and } z \in Z\}$.

Two discrete points $X(x_1,...,x_n)$ and $Y(y_1,...,y_n)$ are two *k-neighbours* if $|x_i - y_i| \leq 1$ for $i \in [1,n]$ and $k \leq n - \sum_{i=1}^n |x_i - y_i|$. Two discrete points are *k-adjacent* if they are k-neighbours but not equal. The proposed notation is adapted to extensions to higher dimensions. A *k-path* in a discrete object A is a sequence of discrete points all in A such that consecutive pairs of points are k-neighbours. A discrete object A is *k-connected* if there is a k-path between two arbitrary points in A. For B a subset of a discrete object A, B is *k-separating* in A if $A \backslash B$ is not k-connected. A discrete object is said to be k-separating if it is k-separating in $\mathbf{Z}^n$. Let A be a k-separating discrete object such that $\overline{A}$ has exactly two k-connected components. A *k-simple* point in A is a discrete point p such that $A \backslash p$ is k-separating. A *simple* point in A is a discrete point that is a k-simple point for some k.

Let's consider three different Euclidean points A, B, and C. The Euclidean line containing A and B, is noted AB. The Euclidean line segment joining A to B is noted (AB). The Euclidean 3D plane containing A, B, and C is noted ABC. The Euclidean triangle with vertices A, B and C is noted (ABC).

A discrete straight line and a discrete plane have been analytically defined by J.-P. Reveillès in the discrete domain as solutions of a double Diophantine inequation [Reve91]:

A discrete 2D line is defined by: $L(d,a,b,\omega) = \left\{(x,y) \in \mathbf{Z}^2 \mid 0 \leq ax + by + d < \omega\right\}$:

A discrete 3D plane is defined by: $P(d,a,b,c,\omega) = \left\{(x,y,z) \in \mathbf{Z}^3 \mid 0 \leq ax + by + cz + d < \omega\right\}$

where $a,b,c,d \in \mathbf{Z}^n$, $\omega \in \mathbf{N}^*$, $gcd(a,b)=1$ for the 2D line and $gcd(a,b,c)=1$ for the 3D plane.

The *coefficients* of the discrete analytical line (resp. plane) (and also the *components of the normal vector*) are a, b (and c). ω is called the *arithmetical thickness* and d the *translation constant* of the line (plane). $\Pi(L,X)=ax+by+d$ and $\Pi(P,X)=ax+by+cz+d$ are called the *control value* of L and P respectively in X.

We can suppose without loss of generality that $gcd(a,b)=1$ for a 2D discrete line because if $gcd(a,b)=u>1$ then $L(d,a,b,\omega) = L\left(\lfloor \frac{d}{u} \rfloor, \frac{a}{u}, \frac{b}{u}, \omega'\right)$ where $\omega' = \lfloor \frac{\omega}{u} \rfloor + 1$ if $\left\{\frac{\omega}{u}\right\} > \left\{\frac{d}{u}\right\}$ and $\omega' = \lfloor \frac{\omega}{u} \rfloor$ else. An equivalent result is verified for the discrete 3D plane.

A discrete object is said to be *k-tunnel free* if a k-path can "go through" the discrete object without intersection. In our case, a discrete analytical 2D line $T=L(d,a,b,\omega)$ (resp. 3D plane $T=P(d,a,b,c,\omega)$) has a k-tunnel if there exists two k-neighbour discrete points A and B satisfying $\Pi(T,A)<0$ and $\Pi(T,B) \geq \omega$. A discrete object that is 0-tunnel free is said to be *tunnel free*.

A 2D discrete line $L(d,a,b,\omega)$ is tunnel free if and only if $\omega \geq |a|+|b|$; it is tunnel free and without simple points if and only if $\omega = |a|+|b|$ [Reve91]. A 3D discrete plane $P(d,a,b,c,\omega)$ is tunnel free if and only if $\omega \geq |a|+|b|+|c|$; it is tunnel free and without simple points if and only if $\omega=|a|+|b|+|c|$ [Andr92] [AS95].

Box1d(A,B), where $A(x_A)$ and $B(x_B)$ are 1D Euclidean points, is the Euclidean interval $[min(x_A, x_B), max(x_A, x_B)]$. *Box2d(A,B)*, where $A(x_A,y_A)$ and $B(x_B,y_B)$ are 2D Euclidean points, is the Euclidean rectangle $Box2d(A,B) = Box1d(A_x, B_x) \times Box1d(A_y,B_y)$. *Box3d(A,B)*, where $A(x_A,y_A,z_A)$ and $B(x_B,y_B,z_B)$ are 3D Euclidean points, is the Euclidean box defined by $Box3d(A,B) = Box1d(A_x,B_x) \times Box1d(A_y,B_y) \times Box1d(A_z,B_z)$.

A 2D discrete point (3D discrete point) in a 1-path (resp. 2-path) L is called a *Jordan point* if and only if it has exactly two 1-adjacent (resp. 2-adjacent) points in L. A 1-path (resp. 2-path) L in 2D (resp. 3D) is called a *Jordan curve* (also called by some authors a *simple* curve) if and only if each point of L is a Jordan point. An *umbrella* at a point Y in a 2D 1-path (resp. 3D 2-path) is a triple (X,Y,Z), where X and Z are the two 1-adjacent (resp. 2-adjacent) points to Y in the path.

An *umbrella* at a point Y in a 3D 2-connected plane is a circular permutation (up to its orientation) of unit squares incident to Y, included in the plane, and such that (1) two consecutive squares in the permutation share a common edge, and (2) only two squares share a common edge. A point in a 2-connected discrete plane P is called a *Jordan point* if and only if it has exactly one umbrella incident to it. If any point of a 2-connected plane is a Jordan point then the plane is a 2-manifold [Fra95b] [Fra95a].

3. Supercover Primitives: Properties and Algorithms

In this section, we are going to examine the properties of several 1D, 2D, and 3D Supercover primitives, and provide algorithms for them. The main mathematical results are the theorems. A Supercover primitive can have two main types of algorithms. The first type is a generation algorithm that generates all the discrete points of the Supercover of a Euclidean primitive. The second type is a localization algorithm that tells us if a given discrete point belongs or not to the Supercover of a Euclidean primitive. While all the primitives have an associated localization algorithm, only the finite sized primitives have an associated generation algorithm. An example for this is the 2D line. We propose a generation and localization algorithm for the Supercover of a 2D line segment joining two points but only a localization algorithm for the Supercover of a complete 2D line. Note that by brute-force, it is always possible to generate the Supercover of a primitive by localizing all its points in a given finite sized region.

This section is divided in three subsections corresponding to the Supercovers of boxes and points, and for 2D and 3D primitives. We don't give all the details of the algorithms since it would be too long but the complete algorithms are available, in a C-language source file, at the *Supercover Home Page* :

$$\text{http://dpt-info.u-strasbg.fr/~nehlig/Supercover.html}$$

The aim of this feature is to allow easy access to a basic Discrete Geometry Library for comparisons, tests and developments. Future revisions will include primitives for more complex objects, extensions on higher dimensions as well as *bubble-free* Supercovers.

3.1 Supercover Boxes and Points

Let's consider for what follows the 3D rational points $A(x_A/R, y_A/R, z_A/R)$, $B(x_B/R, y_B/R, z_B/R)$, and $C(x_B/R, y_B/R, z_B/R)$, *with* x_A, y_A, z_A, x_B, y_B, z_B, $x_C, y_C, z_C \in \mathbf{Z}^3$ and $R \in \mathbf{N}^*$. Let's consider also the 3D discrete point $Q(x_Q, y_Q, z_Q)$, $x_Q, y_Q, z_Q \in \mathbf{Z}$. This point is used in the following localization algorithms.

Proposition 1: *Supercover Boxes:*

1D box: $(A_x B_x)^\omega = Box1d(A_x, B_x)^\omega = \left[\left\lceil \dfrac{2\min(x_A, x_B) + R - 1}{2R} \right\rceil, \left\lfloor \dfrac{2\max(x_A, x_B) + R}{2R} \right\rfloor \right].$

2D box: $Box2d(A_{xy}, B_{xy})^\omega = Box1d(A_x, B_x)^\omega \times Box1d(A_y, B_y)^\omega.$

3D box: $Box3d(A, B)^\omega = Box2d(A_{xy}, B_{xy})^\omega \times Box1d(A_z, B_z)^\omega.$

2D triangle bounding box:

$\quad Box\_triang2d(A_{xy} B_{xy} C_{xy})^\omega = Box2d(A, B)^\omega \cup Box2d(B, C)^\omega \cup Box2d(C, A)^\omega$

$\quad Box\_triang2d(A_{xy} B_{xy} C_{xy})^\omega = Box1d(min(A_x, B_x, C_x), max(A_x, A_y, A_z))^\omega$

$\qquad \times Box1d(min(A_y, B_y, C_y), max(A_y, B_y, C_y))^\omega.$

3D triangle bounding box:

$\quad Box\_triang3d(ABC)^\omega = Box3d(A, B)^\omega \cup Box3d(B, C)^\omega \cup Box3d(C, A)^\omega$

$\quad Box\_triang3d(ABC)^\omega = Box\_triang2d(A_{xy}, B_{xy}, C_{xy})^\omega$

$\qquad \times Box1d(min(A_z, B_z, C_z), max(A_z, B_z, C_z))^\omega.$

The Supercover of a discrete point is simply a particular case of an interval, therefore :

Proposition 2:   *Supercover of a 1D point A:* $A^\omega = Box1d(A,A)^\omega$.

                  Supercover of a 2D point: $A^\omega = Box2d(A,A)^\omega$.

                  Supercover of a 3D point: $A^\omega = Box3d(A,A)^\omega$.

The Supercover of a 1D (resp. 2D, 3D) point is composed of 1 or 2 (resp. 1 or 2 or 4. 1 or 2 or 4 or 8) discrete points. A localization algorithm and a generation algorithm for points are immediately deduced.

3.2 Dimension 2

In dimension 2, we study the supercovers of a 2D line defined by two rational points, a 2D line segment joining two rational points and a 2D triangle defined by three rational points. Let's consider for what follows the 2D rational points $A(x_A/R, y_A/R)$, $B(x_B/R, y_B/R)$, and $C(x_C/R, y_C/R)$, with $x_A, y_A, x_B, y_B, x_C, y_C \in Z$ and $R \in N^*$. Let's consider also the 2D discrete point $Q(x_Q, y_Q)$, with $x_Q, y_Q \in Z$. This point is used in the following localization algorithms.

3.2.1 2D Supercover line

Let's consider the two points A and B and define $dx = x_B - x_A$ and $dy = y_B - y_A$.

If $dx = dy = 0$ it means that $A = B$ and therefore $AB^\omega = A^\omega$. We'll assume for what follows that A and B are two different points.

Let's define: $a = \dfrac{dy}{GCD(dx, dy)}$ and $b = \dfrac{-dx}{GCD(dx, dy)}$ and $d = -ax_A - by_A$.

The points A and B belong to the Euclidean line $L: ax + by + d/R = 0$. For what follows we

call $\omega_1 = -\left\lfloor \dfrac{R(|a|+|b|)+2d}{2R} \right\rfloor$ and $\omega_2 = \left\lfloor \dfrac{R(|a|+|b|)-2d}{2R} \right\rfloor$. The Supercover of $L = AB$ is

given by:

Theorem 3: *Supercover of a 2D straight line:*

$$AB^\omega = \left\{ (x,y) \in Z^2 \,\middle|\, \omega_1 \leq ax + by \leq \omega_2 \right\} = L(-\omega_1, a, b, \omega_2 - \omega_1 + 1) . \quad \bullet$$

The Supercover AB^ω is a Reveillès 2D discrete line of arithmetical thickness $\omega = \omega_2 + 1 - \omega_1$. When $d/R = k$, an integer, and $|a| + |b|$ even or $d/R = k + 1/2$ and $|a| + |b|$ odd then $\omega = |a| + |b| + 1$. The discrete line is then 1-connected with simple points located at *4-bubbles*. In all the other cases, $\omega = |a| + |b|$ and the line is a 1-connected Jordan line, that is without *simple* points; these lines are called *standard* lines. Note that, if $R = 1$, that is, if A and B are integer points, then $\omega_1 = -\omega_2$; moreover, the control values in A and B are 0.

A *4-bubble* in a 2D (resp. 3D) discrete line is a set of four different integer points P_1, P_2, P_3, P_4 where P_1 is a 1-adjacent (resp. 2-adjacent) of P_2, P_2 of P_3, P_3 of P_4 and P_4 of P_1.

Ex. In the Supercover of the Euclidean 2D line $x + y = 0$ any point belongs to a 4-bubble. The same holds for the line $y = 1/2$, and for the line $x + 2y + 1/2 = 0$.

Proposition 4: *Location of 4-bubbles:*

Let's suppose that the equation of the Supercover AB^ω is: $\omega_1 \leq ax+by \leq \omega_2$, with $0 \leq a \leq b$. The Supercover AB^ω has 4-bubbles if and only if $\omega = \omega_2 + 1 - \omega_1 = a+b+1$. The control values at the points of a 4-bubble are ω_1, ω_2, $\omega_2 - a$ or $\omega_1 + a$. The simple points are the points of control value ω_1 and ω_2. •

Sketch of proof: According to [Reve91], a discrete line is 4-connected, without simple points, if and only if its arithmetical thickness is $a+b$; a Supercover line has thus a bubble if and only if its arithmetical thickness is $a+b+1$. Let r be the remainder at a point with minimum x coordinate and maximum y coordinate, belonging to a given bubble. The remainders at the other three points of the bubble are then $r+a$, $r+b$, and $r+a+b$, and these 4 numbers have to belong to the $[\omega_1 , \omega_2]$ interval. The only solution is $r = \omega_1$, and $r+a+b = \omega_2$. The last part of the proposition can similarly be obtained.

Proposition 5: Let L be a Supercover 2D line, and let p a point of L. If p does not belong to a 4-bubble, then p is a Jordan point. •

A line is not of finite size, therefore we'll only propose a localization algorithm for this primitive. A localization algorithm Super_local_line2d(A,B,Q) can be deduced immediately from theorem 3.

3.2.2 2D Supercover line segment

It is easy to see that the Supercover of a Euclidean 2D line segment joining two rational points A and B is given by:

Proposition 6: *Supercover of a 2D discrete line segment:*

$$(AB)^\omega = AB^\omega \cap Box2d(A,B)^\omega. \bullet$$

A localization algorithm Super_local_lineseg2d(A,B,Q) can be immediately deduced.

The design of an efficient generation algorithm Super_gen_lineseg2d(A,B) is however more complicated. Let's consider $dx = x_B-x_A$, $dy = y_B-y_A$,used in the computation of the coefficients of the discrete line AB^ω. If dx or dy are null then it means that $A=B$ or that we have a horizontal or a vertical line. All three cases can be handled by Super_gen_box2d(A,B).

Now, let's suppose that $0 \leq dy \leq dx$. This can be obtained by symmetry. The first thing we do is to compute the coefficients a,b,ω_1,ω_2 of the 2D Supercover line according to theorem 3 and the bounding box $Box2d(A,B)^\omega=[x_d,x_f] \times [y_d,y_f]$. We can now use an incremental generation algorithm where (x_d,y_d) will be the first point to be generated and (x_f,y_f) the last. The control value Π of the incremental algorithm is initialized with $\Pi=ax_d+by_d$. At each loop of the algorithm we advance x and adjust P by an increment of a. As long as we have $P \leq \omega_2$, we continue. Once this is not true anymore, we have to increment y. In order to detect if there is a 4-bubble, we decrement then x by 2. If this point belongs to the line then there was a 2-bubble at that point and we generate it, else we simply continue by incrementing x in the regular loop. The algorithms stops when it reaches the point (x_f,y_f).

3.2.3 Supercover 2D triangle

Let's call T the triangle $T=(ABC)$. Let's first assume that the three points are non aligned because else we have simply $T^\omega = (AB)^\omega \cup (BC)^\omega \cup (CA)^\omega$.

Let's assume the Cartesian equation of the Euclidean line AB is $a_1x+b_1y+d_1/R=0$ defined so that $a_1x_C+b_1y_C+d_1/R >0$. In the same way, the equation of BC is $a_2x+b_2y+d_2/R =0$ defined so that $a_2x_A+b_2y_A+d_2/R >0$ and the equation of CA is $a_3x+b_3y+d_3/R =0$ defined so that $a_3x_B+b_3y_B+d_3/R >0$.

Let's consider now the Supercovers of these three lines:

AB^ω: $\omega_1 \le a_1x+b_1y \le \rho_1$; BC^ω: $\omega_2 \le a_2x+b_2y \le \rho_2$ and CA^ω: $\omega_3 \le a_3x+b_3y \le \rho_3$.

Let's call H_k the discrete half space $a_kx+b_ky \ge \omega_k$ and I_k the discrete half space $a_kx+b_ky > \rho_k$, $k=1,2,3$, then:

Theorem 7: *Supercover of a 2D rational triangle:*

$$T^\omega = (ABC)^\omega = H_1 \cap H_2 \cap H_3 \cap \text{Box\_triang2d}(ABC^\omega)$$

$$T^\omega = (ABC)^\omega = ((AB)^\omega \cup (BC)^\omega \cup (CA)^\omega \cup (I_1 \cap I_2 \cap I_3)) \cap \text{Box\_triang2d}(ABC)^\omega. \bullet$$

We call *interior* of T^ω the integer set $I_1 \cap I_2 \cap I_3$; it can be empty. We call *boundary* of T^ω the set $((AB)^\omega \cup (BC)^\omega \cup (CA)^\omega) \cap \text{Box\_triang2d}(ABC)^\omega$. Thus, T^ω is the union of its interior and its boundary. Moreover, the interior and the boundary of T^ω do not intersect.

Note that the intersection with the Supercover of the bounding box $\text{Box\_triang2d}(ABC)^\omega$ is necessary to define the boundary of T^ω, because the intersection of $(AB)^\omega$ and $(BC)^\omega$, for example, can contain points outside T^ω, as shown on the following example: AB is the line $y = 0$, AC is $x - 9y = 0$, and $A = (0, 0)$, B and C have positive abscissa; the point $(-1, 0)$ belongs to $AB^\omega \cap AC^\omega$ but not to T^ω.

A localization algorithm $\text{Super\_local\_triang2d}(A,B,C,Q)$ is simple to design with the results of proposition 6 and theorem 7.

The design of a generation algorithm is slightly more complicated. We can of course generate the triangle by brute-force by testing what points of the Supercover of the bounding box belong to the Supercover triangle with help of the localization algorithm. This is however not efficient. A more efficient method consists in an adaptation of the classical scanline filling algorithm. We won't describe here the algorithm which would be too long. Let's just remark that if we have a discrete line L: $\omega_1 \le a_1x+b_1y \le \rho_1$, with $a>0$, then for a given $y=y_i$ we have $-\left\lfloor \dfrac{by_i - \omega_1}{a} \right\rfloor \le x \le \left\lfloor \dfrac{\omega_2 - by_i}{a} \right\rfloor$. By doing that for two lines it is easy to determine a span of pixels to fill up.

3.3 Dimension 3

In dimension 3, we study the supercovers of a 2D line defined by two 2D rational points, a 3D line segment joining two rational 3D points, a 3D plane defined by three 3D rational points and a 3D triangle defined by three rational 3D points. Let's consider for what follows the 3D rational points $A(x_A/R, y_A/R, z_A/R)$, $B(x_B/R, y_B/R, z_B/R)$, and $C(x_B/R, y_B/R, z_B/R)$, with x_A, y_A, z_A, x_B, y_B, z_B, $x_C, y_C, z_C \in \mathbf{Z}^3$ and $R \in \mathbf{N}^*$. Let's consider also the 3D discrete point $Q(x_Q, y_Q, z_Q)$, $x_Q, y_Q, z_Q \in \mathbf{Z}$. This point is used in the following localization algorithms.

3.3.1 Supercover of a 3D straight line

Let's consider the two points A and B and define $dx = x_B-x_A$, $dy = y_B-y_A$ and $dz = z_B-z_A$.

If $dx=dy=dz=0$ it means that $A=B$ and therefore $AB^\omega=A^\omega$. We'll assume for what follows that A and B are two different points.

Theorem 8: *Supercover of a 3D straight line:*

$$AB^\omega = (A_{xy}B_{xy}{}^\omega)^z \cap (A_{xz}B_{xz}{}^\omega)^y \cap (A_{yz}B_{yz}{}^\omega)^x$$

$$AB^\omega = P(-\omega_1,a_1,b_1,0,\rho_1\text{-}\omega_1+1) \cap P(-\omega_2,a_2,0,b_2,\rho_2\text{-}\omega_2+1) \cap P(-\omega_3,0,a_3,b_3,\rho_3\text{-}\omega_3+1)$$

where $A_{xy}B_{xy}{}^\omega = L(-\omega_1,a_1,b_1,\rho_1\text{-}\omega_1+1)$; $A_{xz}B_{xz}{}^\omega = L(-\omega_2,a_2,b_2,\rho_2\text{-}\omega_2+1)$

and $A_{yz}B_{yz}{}^\omega = L(-\omega_3,a_3,b_3,\rho_3\text{-}\omega_3+1)$. •

Sketch of proof: 1) $(A_{xy}B_{xy}{}^\omega)^z \cap (A_{xz}B_{xz}{}^\omega)^y \cap (A_{yz}B_{yz}{}^\omega)^x \supset Ab^\omega$. This part is obvious. 2) $Ab^\omega \supset (A_{xy}B_{xy}{}^\omega)^z \cap (A_{xz}B_{xz}{}^\omega)^y \cap (A_{yz}B_{yz}{}^\omega)^x$. This inclusion is due to following lemma: Let D be a Euclidean 3D line and let (u,v,w) be an integer coordinate point such as the projection of D on the (x,y) (resp. (y,z), (z,x)) plane intersects the pixel of centre (u,v) (resp. (v,w), (w,u)) of this plane, then D intersects the voxel of centre (u,v,w). •

The analytical definition of the Supercover 3D lines is based on three 2D lines (respectively in the planes xy, yz and zx) or three planes. It can be proved that two of them are, generally, not enough, as it can be seen in Figure 1. This is a major difference with classical definitions. The fact that the intersection of three discrete planes defines a discrete line creates a symmetry in the 3D line that does not exist in classical definitions.

It is easy to see that a 3D Supercover line is 2-connected; it can be a 2-connected Jordan curve (that is without a simple point); but it can have simple points if it contains 4-bubbles, like a 2D Supercover line, or 8-bubbles. An *8-bubble* in a 3D line or plane is a set of eight discrete points forming a $1 \times 1 \times 1$ unit cube. A 8-bubble in a 3D segment is visible on the Figure 2. In the same Figure and in Figure 1, a 4-bubble in a 3D segment is also visible.

A line is not of finite size, therefore we'll only propose a localization algorithm for this primitive. A localization algorithm Super_local_line2d(A,B,Q) can be deduced immediately from theorem 8.

3.3.2 Supercover of a 3D straight line segment

It is easy to see that the Supercover of a Euclidean 3D line segment joining two rational points A and B is given by

Proposition 9: *Supercover of a 3D line segment:*

$$(AB)^\omega = AB^\omega \cap Box3d(A,B)^\omega. •$$

Figure 1 shows an example of a Supercover 3D segment.

Proposition 10: *Supercover of a 2D line segment in 3D space:*

Let's suppose that $z_B\text{-}z_A=0$; then $(AB)^\omega = (A_{xy}B_{xy})^\omega \times Box1d(A_z,B_z)^\omega$. •

A localization algorithm Super_local_lineseg2d(A,B,Q) can be immediately deduced. The design of an efficient generation algorithm Super_gen_lineseg3d(A,B) is, however, more complicated. Let's consider $dx = x_B\text{-}x_A, dy = y_B\text{-}y_A$, and $dz = z_B\text{-}z_A$. If two or three of the three dx, dy and dz are null then it means that $A=B$ or that we have a line parallel to an axis. All these cases can be handled by Super_gen_box3d(A,B). Let's suppose that only one is equal to zero, $dz=0$ for instance. We generate a 2D segment and apply proposition 6.

Now, let's suppose that $0 \le dz \le dy \le dx$. We know from theorem 8 that a point (x,y,z) belongs to the 3D line if (x,y) belongs to $A_{xy}B_{xy}{}^\omega$, if (y,z) belongs to $A_{yz}B_{yz}{}^\omega$ and if (x,z) belongs to $A_{xz}B_{xz}{}^\omega$. What we do therefore is to run three 2D line segment generation algorithms together. At each step we verify that all three conditions are verified.

3.3.3 Supercover 3D plane

Let's consider the three rational 3D points A, B, and C.

Let's define $dx=(y_B-y_A)(z_C-z_A)-(y_C-y_A)(z_B-z_A)$, $dy=(z_B-z_A)(x_C-x_A)-(z_C-z_A)(x_B-x_A)$, and $dz=(x_B-x_A)(y_C-y_A)-(x_C-x_A)(y_B-y_A)$.

If $dx=dy=dz=0$ then it means that AB and AC are collinear and do not define a 3D plane but a 3D line or a 3D point. We'll assume for what follows that A, B and C are three non aligned points.

Let's define: $a = \dfrac{dx}{GCD(dx,dy,dz)}$, $b = \dfrac{dy}{GCD(dx,dy,dz)}$, $c = \dfrac{dz}{GCD(dx,dy,dz)}$ and $d = -ax_A-by_A-cz_A$. The points A, B and C belong to the Euclidean plane P: $ax+by+cz+d/R=0$.

We note $\omega_1=-\left\lfloor \dfrac{R(|a|+|b|+|c|)+2d}{2R} \right\rfloor$ and $\omega_2=\left\lfloor \dfrac{R(|a|+|b|+|c|)-2d}{2R} \right\rfloor$.

The Supercover of $P=ABC$ is then given by:

Theorem 11: *Supercover of a 3D plane:*

$P^\omega = ABC^\omega = \left\{(x,y,z) \in \mathbf{Z}^3 \middle| \omega_1 \le ax+by+cz \le \omega_2\right\} = P(-\omega_1,a,b,c,\omega_2-\omega_1+1)$. •

The Supercover P^ω is a Reveillès 3D discrete plane of arithmetical thickness $\omega =\omega_2+1-\omega_1$. When $d/R=k$, k integer, and $|a|+|b|+|c|$ even or $d/R=k+1/2$, k integer, and $|a|+|b|+|c|$ odd then $\omega=|a|+|b|+|c|+1$. The discrete plane is 2-connected, tunnel free and it has 8-bubbles. In all the other cases, $\omega=|a|+|b|+|c|$. The discrete plane is then 2-connected, tunnel free, and without simple points (it is a 2-manifold [Fra95b] or [Fra95a]). Note that, if $R=1$, that is, if A, B and C are integer points, then $\omega_1=-\omega_2$; moreover, the control values in A, B and C are 0.

Proposition 12: *Control value at a 8-bubble:*

Let's suppose that the equation of the Supercover P^ω is: $\omega_1 \le ax+by+cz \le \omega_2$, with $0 \le a \le b \le c$. The Supercover P^ω contains 8-bubbles if and only if $\omega=\omega_2+1-\omega_1 = a+b+c+1$. The control values at the points of a 8-bubble are ω_1, ω_1+a, ω_1+b, ω_1+a+b, ω_1+c, ω_1+c+a, ω_1+c+b, $\omega_1+a+b+c = \omega_2$. •

Proposition 13: Let P be a Supercover plane, and let p a point of P. If p does not belong to a 8-bubble then p is a Jordan point. •

Because the intersection of two non parallel Euclidean planes is a Euclidean line we have:

Theorem 14: *Intersection of 3D Supercover planes:*

Let's consider two 3D rational points A and B. The intersection of the supercovers of all the Euclidean planes containing AB is equal to AB^ω. •

Since a 3D plane is not finite, we propose only a localization algorithm called Super_local_plane3d(A,B,C,Q). With help of theorem 11 the algorithm design is straight forward.

3.3.4 Supercover 3D triangle

Let's call T the triangle $T=(ABC)$. Let's first assume that the three points are non aligned because else we have simply $T^\omega = (AB)^\omega \cup (BC)^\omega \cup (CA)^\omega$.

Theorem 15: *Supercover of a 3D triangle:*

$$T^\omega = (ABC)^\omega = (T_{xy}^\omega)^z \cap (T_{xz}^\omega)^y \cap (T_{yz}^\omega)^x \cap ABC^\omega. \bullet$$

A Supercover 3D triangle is shown in Figure 2.

Note that, like in theorem 8, two of the three triangles T_{xy}, T_{xz}, T_{yz} are not enough, as it can be shown on examples. Furthermore, as any polygon can be tiled using triangles, the Supercover of a polygon is defined as the union of the Supercovers of a tiling set of triangles.

A localization algorithm Super_local_triang3d(A,B,C,Q) is easy to design with theorem 15. The generation algorithm Super_gen_triang3d(A,B,C) of a 3D triangle is actually much easier to implement than the 2D algorithm. After having dealt with the degenerate cases and so being sure that the three points are not aligned we can compute the coefficients of the Supercover of the Euclidean plane P=ABC.

Let's suppose that the Supercover of P is equal to P^ω: $\omega_1 \le ax+by+cz \le \omega_2$. Let's suppose here, without loss of generality that $c>0$. Now we generate the 2D Supercover triangle T_{xy}^ω and for each (x_i, y_i) belonging to T_{xy}^ω we compute all the corresponding (x_i, y_i, z_k) belonging to P^ω. We know that $-\left\lfloor \dfrac{ax_i + by_i - \omega_1}{c} \right\rfloor \le z_k \le \left\lfloor \dfrac{\omega_2 - ax_i - by_i}{c} \right\rfloor$. For each (x_i, y_i, z_k) we do a localization Super_local_triang2d(A_{yz}, B_{yz}, C_{yz}, (y_i, z_k))) and a localization Super_local_triang2d(A_{xz}, B_{xz}, C_{xz}, (x_i, z_k))). We generate(x_i, y_i, z_k) if and only if the point (y_i, z_k) belongs to T_{yz}^ω and if (x_i, z_k) belongs to T_{xz}^ω.

4. Conclusion

We have defined the Supercover of continuous objects. We have then found out analytical description, by the way of inequalities (sets of Diophantine equations), of the basic primitives of the geometry: 2D straight lines, segments and triangles, 3D straight lines, planes and triangles. We have given geometric properties concerning the arithmetical thickness of the lines and of the planes, and the bubbles. Basic algorithms for localization and for generation have been given; they are adaptations of classical algorithms for continuous primitives.

With these primitives and algorithms a *discrete geometric modelling* whose primitives are polygons and polyhedra is possible; the first results are shown in [ANF97].

New discrete geometric problems are in order. The main one is: is it possible to use discrete 2D straight lines, and discrete planes, whose thickness is lesser than that of Supercovers, and verifying analogous theorems, especially an analogue of theorem 14? Examples show that it is not the case of naive planes of [Reve91] and naive lines of [FR95].

References

[Andr92]   E. Andres, *Cercles discrets et Rotations discretes* [Discrete Cercles and Discrete Rotations], Ph.D. Thesis, Université Louis Pasteur, Strasbourg (France). Dec. 1992 (in French).

[AS95]   E. Andres, and C. Sibata, *Discrete hyperplanes in arbitrary dimensions*. to be published in CVGIP-GMIP.

[ASA96]   E. Andres, C. Sibata, and R. Acharya, *Supercover 3D Polygon*. Proc. Int. Workshop on Discrete Geometry for Computer Imagery, LNCS, vol.1176, pp. 237-242.

[ANF97]   E. Andres, Ph. Nehlig, and J. Françon, *Tunnel-free Supercover 3D polygons and polyhedra*, Eurographics'97, BUDAPEST, Hungary, Sept 4-8 (1997). Computer Graphics Forum (Blackwell Publishers), vol 16(3), Conference issue, pp. C3-C13.

[Cohe95]   D. Cohen and A. Kaufman. *Fundamentals of surface voxelization*. CVGIP-GMIP, 57(6), Nov.95, pp.453-461.

[Fra95a]   J. Françon, *Arithmetic planes and combinatorial manifolds*, 5th International Conference Discrete Geometry in Computer Imagery, Clermont-Ferrand, 25-27 Sept. 1995, pp. 209-217.

[Fra95b]   J. Françon, *Sur la topologie d'un plan arithmétique* [About the topology of the arithmetical plane], Theor. Comp. Sc. 156, 1996, pp. 159-176 (in French).

[FR95]   O. Figueiredo, and J.-P. Reveillès, *A contribution to 3D digital lines*, 5th International Conference Discrete Geometry in Computer Imagery, Clermont-Ferrand, 25-27 Sept. 1995, pp. 187-198.

[Reve91] J.-P. Reveillès, *Géométrie Discrète, calcul en nombres entiers et algorithmique* [Discrete Geometry, Integer Number Calculus, and Algorithmics], Thèse d'état. Université Louis Pasteur, Strasbourg (France), Dec. 1991 (in French).

254

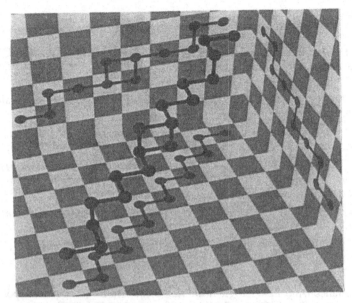

Figure 1: The Supercover of a 3D segment

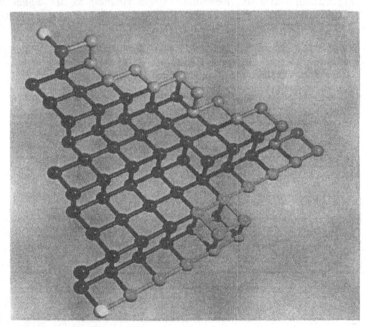

Figure 2: The Supercover of a 3D triangle
*The vertices are of one color, the edges are each of a different color,
the interior is of another color*

Author Index

Springer
and the
environment

At Springer we firmly believe that an international science publisher has a special obligation to the environment, and our corporate policies consistently reflect this conviction.

We also expect our business partners – paper mills, printers, packaging manufacturers, etc. – to commit themselves to using materials and production processes that do not harm the environment. The paper in this book is made from low- or no-chlorine pulp and is acid free, in conformance with international standards for paper permanency.

Springer

Lecture Notes in Computer Science

For information about Vols. 1–1272

please contact your bookseller or Springer-Verlag